The breathtaking number of mergers and joint ventures among agribusiness firms has left independent American farmers facing the power of an increasingly concentrated buying sector. The origin of farmers' concern with such economic concentration dates back to protests against meatpackers and railroads in the late nineteenth century. Jon Lauck examines the dimensions of this problem in the American Midwest in the decades following World War II. He analyzes the nature of competition within meat-packing and grain markets. In addition, he addresses concerns about corporate entry into production agriculture and the potential displacement of a production system defined by independent family farms.

Lauck also considers the ability of farmers to organize in order to counter the market power of large-scale agribusiness buyers. He explores the use of farmer cooperatives and other mechanisms which may increase the bargaining power of farmers. The book offers the first serious historical examination of the National Farmers Organization, which fully embraced the bargaining power cause in the postwar period. Lauck finds that independent farmers' attempts at organization have been more successful than previously recognized, but he also shows that their successes have been undermined by the growing concentration and power of agribusiness firms, justifying a new approach to antitrust law in agricultural markets.

Jon Lauck is editor-in-chief of the *Minnesota Journal of Global Trade*.

University of Nebraska Press
Lincoln NE 68588-0484

www.nebraskapress.unl.edu

JON LAUCK

American Agriculture and the Problem of Monopoly

THE POLITICAL ECONOMY OF GRAIN BELT FARMING, 1953–1980

University of Nebraska Press, Lincoln and London

FOR MOM AND DAD,
PILLARS OF THE REPUBLIC

Excerpts from Linda Hasselstrom's poem "Coffee Cup Cafe" are used with permission from *Land Circle: Writings Collected from the Land*, by Linda Hasselstrom. © 1991. Fulcrum Publishing, Inc., Golden, Colorado, USA. All rights reserved.
Acknowledgments of previously published work appear on p. xiv.
© 2000 by the University of Nebraska Press
All rights reserved
Manufactured in the United States of America

∞

Library of Congress Cataloging-in-Publication Data
Lauck, Jon, 1971–
American agriculture and the problem of monopoly: the political economy of grain belt farming, 1953–1980 / Jon Lauck.
    p.   cm.
Includes bibliographical references and index.
ISBN 0-8032-2932-1 (cl.: alk. paper)
1. Agriculture—Economic aspects—Middle West—History—20th century. 2. Agriculture and state—Middle West—History—20th century. 3. Agricultural industries—Mergers—Middle West—History—20th century. I. Title.
    HD1773.A3L38   2000
338.1'0977'09045—dc21   99-38710
                CIP

CONTENTS

	List of Tables	vii
	Preface	ix
1	The Problem	1
2	The Corporate Farming Debate	19
3	The Political Economy of Meatpacking and Grain Processing	39
4	The Grain-Trading "Cartel"	62
5	The NFO and Farm Bargaining	84
6	Farmer Cooperative Marketing	109
7	The State and Agricultural Organization	136
	Conclusion	163
	Epilogue: Toward an Agrarian Antitrust	177
	Notes	183
	Index	255

TABLES

Table 1	Changes in Farm Structure in the Grain Belt, 1950–1992	8
Table 2	The Diversification of Armour and Company	41
Table 3	Declining Consumption and Prices of Meat, 1970–1989	47
Table 4a	Estimates of Cost of Beef-Slaughtering Plants of Varying Capacity (in thousands of dollars)	49
Table 4b	Estimates of Cost of Hog-Slaughtering Plants of Varying Capacity (in thousands of dollars)	49
Table 5a	Total Sales of Big Four (in millions of dollars)	51
Table 5b	Total Sales of Little Five (in millions of dollars)	51
Table 6	Packing Industry Master Agreement Wages versus IBP Wages (in dollars per hour)	55
Table 7	International Wheat Agreement, 1953	67
Table 8	Number of Firms Exporting	82

PREFACE

My family moved off the farm before the worst years of the 1980s agricultural depression. As the descendants of farmers in a farming community, and not far removed from the farm ourselves, we traveled to Pierre, South Dakota, for the big farm rallies at the capitol, hoping to be of some help. After one of the rallies, the governor sent the entire state legislature to Washington to lobby for aid to farmers, a measure of the gravity of the crisis. In many ways the pressures facing farmers in the 1980s were more wrenching than, say, the 1890s, since many farmers operated farms owned by their family for decades, some even for a century. The farm crisis gripped our community—Jessica Lange's movie *Country*, which depicted the struggle of a typical Iowa farm family on a century farm, was so depressing that some people had to leave the theater. While there were many explanations for the hard times farmers faced, one that seemed to persist was the alleged abuses of a few powerful economic interests, historically explained as the monopoly problem, the claim that prompted the writing of this book.

Although the focus of this book is the economic dimensions of the monopoly problem, in the process of my research I have been constantly reminded of the noneconomic nature of many farmers' complaints, something especially observable to those who lived through the 1980s farm crisis. In particular, I kept thinking about the classical republican nature of the complaints. As J. G. A. Pocock has noted, "If freehold land is the guarantee of virtue, of moral and political personality, and if corruption is any social change which tends to displace it as the material base of human relations, then the myth of the American frontier, of the farming West . . . takes its place in the history of civic humanism," or republicanism. The letters I read from some farm families in the course of my research, along with my memories of the 1980s, reinforce what Americans have always feared—that the agrarian basis of the republican order was fleeting. Pocock: "From Jefferson to Frederick Jackson Turner and beyond, it was a commonplace that sooner or later the frontier would be closed, the land filled, and the corruptions of history—urbanisation, finance capital, 'the cross of gold,' 'the military-industrial complex'—would overtake America. Here are the origins of American historical pessimism."[1]

The frontier was long since closed in my period of study, but farming continued to be a prominent public issue—at least in the 1950s and early 1960s—and the passing of the agrarian order a fearful development. The disappearance of the American farmer, long thought to be the anchor of the republic, proved especially frightening during the postwar challenge of international communism and what many viewed as the wholesale assault on moral and social institutions in the 1960s. The "historical pessimism" Pocock notes reached fever pitch in rural areas in the post–World War II years, although its memory is faint, since most historians end their attention to farmers with the Great Depression. My time in the archives exposed me to what Christopher Lasch called "the darker voices," farmers and farm advocates who failed to see the developments in American agriculture as anything resembling "progress." Instead, they viewed such developments as a fateful step backward for a republic dependent on civic virtue, decentralized economic institutions, a large class of property owners, and community.[2]

Many farmers, to be sure, accepted the changes in agriculture in the postwar years and thrived using new technologies and greater capital, often bristling at the federal government's economic interventions aimed at solving the "farm problem." These farmers adhered to the powerful American tradition of Lockean liberalism, devoted to property rights, economic freedoms, and civil liberties. Farmers who were less sanguine about the developments in agriculture embraced certain components of republican ideology, devoted to the ideal of dispersed wealth and land, a freeholding citizenry, and the civic virtue and responsibility inherent in small-town and rural culture. Fears about the unraveling of this social arrangement were often expressed in complaints about the "monopoly problem."

Unfortunately, historians and other scholars have given little attention to American farming after World War II. I hope this book can serve as a beginning link between the history of postwar farming and the vigorous debates about the molding of the postwar American political economy, and I hope it will encourage scholars to consider the events of the grain belt when organizing surveys of postwar history, prompting them at least to add important movements such as the National Farmers Organization to their outlines. And by taking seriously the economic context and economic forces that shape social institutions, this work attempts to build on the social history written by the New Rural Historians, whose work is too often cut off from wider economic and political contexts. In so doing, I hope for what Thomas Bender calls a greater "intersection of politics, ideas, and society" in this area of study in the future, and a "return to issues of the state and political democracy."[3]

A look at postwar farming is also part of a much wider conversation about the health of the republic, one that took center stage in the 1960s, about the time farm issues faded from prominence in American politics. Historically, American political culture has been a mixture of the republican and Lockean traditions. Fears that the republican components are fading or are being forgotten are reflected in the criticism of shriveling community ("bowling alone," in Robert Putnam's phrase), collapsing civic institutions and eroding citizenship (the source of "democracy's discontents," in Michael Sandel's phrase), fragmenting cultural traditions (the "disuniting of America," in Arthur Schlesinger Jr.'s phrase), and failing political institutions (the "democratic malaise," in Lasch's phrase).[4] A reintroduction of republican talk and pro-republican public policy, one informed by the importance of farming in this tradition, may help restore the balance. In his speech "The Perpetuation of Our Political Institutions," Lincoln emphasized the importance of remembering the republic's first principles, the "living histories" of the revolutionaries themselves, lest we lose our way, allowing republican virtue to be submerged by narcissism:

> But those histories are gone. They can be read no more forever. They were a fortress of strength; but, what invading foemen could never do, the silent artillery of time has done; the levelling of its walls. They are gone. They were a forest of giant oaks; but the all-resistless hurricane has swept over them, and left only, here and there, a lonely trunk, despoiled of its verdure, shorn of its foliage, unshading and unshaded, to murmur in a few more gentle breezes, and to combat with its mutilated limbs, a few more ruder storms, then to sink, and be no more.

The many "props" the republic had to support it through its first half-century had "decayed, and crumbled away," according to Lincoln, just as the heritage of republicanism and agrarianism has in the 1990s.[5] Michael Sandel notes that "from Aristotle's polis to Jefferson's agrarian ideal, the civic conception of freedom found its home in small and bounded places, largely self-sufficient, inhabited by people whose conditions of life afforded the leisure, learning, and commonality to deliberate well about public concerns. But we do not live that way today." T. S. Eliot observed that liberalism "is a movement not so much defined by its end, as by its starting point; away from, rather than towards something definite." What the American tradition of liberalism is sliding away from is the concern. A more republican and agrarian liberalism, restoring the old balance and holding the Lockean tradition closer to its "starting point," offers a way out of the present crisis, one that appeals to our heritage—what Lincoln called the "mystical chords of memory"—and our better angels, and offers hope.[6]

In the writing of this book I have incurred many debts. In my years at the University of Iowa I had the honor and privilege of working with some world-class scholars. My training started when Ellis Hawley mailed a copy of his syllabus to me in Madison, South Dakota, the summer before I started graduate school—I started reading *The Great War and the Search for a Modern Order* on the back porch as an introduction to my graduate studies. His greatest work, *The New Deal and the Problem of Monopoly*, has also made an obvious impression on this book. Professor Hawley chaired my masters committee and introduced me to the world of historical political economy in his readings course during my first spring in Iowa City. This project started in that seminar, and I thank Professor Hawley for guidance, encouragement, and inspiration. The same spring I served as Deirdre McCloskey's research assistant and was invited to her Sunday Seminar in Economic History. Her books, articles, and comments criticizing neoclassical economics have contributed greatly to my thinking about the noneconomic dimension of antitrust. I thank her for the many original insights, which were extremely helpful in a scholarly environment that at times seems strangely orthodox, and for her many generosities, not the least of which was chairing my dissertation committee. That same eventful year, Colin Gordon published *New Deals* and joined the Iowa faculty, providing me access to some of the latest thinking on modern American political economy. Even though we approach historical problems from different perspectives, Professor Gordon offered many helpful comments on the conceptualization and organization of the project, and I thank him for his weighty contributions. During the fall of that same year, Herb Hovenkamp allowed me to attend his classes in the law school, introducing me to the legal and economic intricacies of antitrust law, and Professor Lawrence Gelfand helped me to publish my first academic article. I thank both of them for their help and efforts on my behalf. Ed Adams, Richard Adelstein, Harold Breimyer, Peter Carstenson, Jim Chen, David Danbom, Gilbert Fite, Dan Gifford, James Giglio, David Hamilton, Kenneth Hendrickson, Chris Kimball, David McGowan, James Matray, Don Muhm, Bill Pratt, Leo Raskind, and Earl Rogers also offered helpful comments along the way. And I also want to extend my sincerest gratitude to E. Thomas Sullivan, Dean of the University of Minnesota Law School. When I was revising my manuscript for publication, Dean Sullivan allowed me to serve as his research assistant while he was revising his antitrust textbook, which was coauthored by Professor Hovenkamp. I thank him for extending me such a unique opportunity, for all that I learned about antitrust law while working with him, and for being such a strong supporter.

I also want to thank the professors at South Dakota State University, where my education into these matters started. In particular, I thank John Miller, Robert Burns, Jerry Sweeney, Michael Funchion, Rodney Bell, Gordon Tolle, Herb Cheever, David Crane, Nels Granholm, John Taylor, and President Robert Wagner. I thank my friends in graduate and law school, especially Jason Duncan, Dave McMahon, Dave Kilroy, Dave Gilbert, Mark Milosch, Chris Moon, and Jennifer Bower for their comments and ongoing conversation about the work of history, and very helpful librarians like Eeva Hoch, Marianne Ryan, Mike O'Sullivan, and Becky Jordan. I would also like to thank Allan Bogue and the two anonymous reviewers for the University of Nebraska Press for their helpful suggestions on the manuscript. And I thank the taxpayers of the great states of South Dakota, Iowa, and Minnesota for generously supporting my education over the years.

Finally, I thank my mom and dad. Their unflinching support, letters, phone calls, and encouragement were more critical than I am able to express. They personify the classical republican tradition—virtuous citizenship, yeoman heritage, defense of the commonwealth. Since they lived much of it, they should be the ones telling the story.

Portions of this book have appeared elsewhere. An early version of chapter 1 appeared as "American Agriculture and the Problem of Monopoly," © 1996 by *Agricultural History*, and is reprinted from *Agricultural History* 70, 2 (1996), by permission. A version of chapter 2 appeared as "The Corporate Farming Debate in the Post–World War II Midwest" in *Great*

Plains Quarterly 18, 2 (spring 1998), and is reprinted by permission of *Great Plains Quarterly*. Portions of chapter 3 are adapted from "Competition in the Grain Belt Meatpacking Sector after World War II" in *The Annals of Iowa* 57 (spring 1998): 135–59, copyright 1998 State Historical Society of Iowa, and is used with the permission of the publisher. Portions of chatper 4 appeared as "Against the Grain: The North Dakota Wheat Pooling Plan and the Liberalization Trend in World Agricultural Markets" in the *Minnesota Journal of Global Trade* 8, 2 (summer 1999). A version of chapter 5 appeared as "The National Farmers Organization and Farmer Bargaining Power" in *Michigan Historical Review* 24, 2 (fall 1998): 88–127. The epilogue reproduces portions of "Toward an Agrarian Antitrust: New Directions in Agricultural Law," published in *North Dakota Law Review Journal* 75, 3 (1999). A version of the conclusion appeared as "The Silent Artillery of Time: The Social Consequences of Economic Change on the Northern Great Plains" in *Great Plains Quarterly* 19, 4 (fall 1999), and is reprinted by permission of *Great Plains Quarterly*.

JON LAUCK
SIOUX FALLS, MAY 1999

I

THE PROBLEM

Farmers identified concentrated economic wealth—the monopoly problem—as a threat to the American republic in the late nineteenth century, and the critique endured. Teddy Roosevelt busted trusts as president, and his distant cousin Franklin condemned "economic royalists" and initiated a full-scale antitrust campaign twenty years later. After World War II the deep sense of urgency about the monopoly problem persisted. John Kenneth Galbraith, for example, argued that "the dominant market of modern capitalism is not one made up of sellers offering either uniform or differentiated products. Rather it is a market of few sellers," and "all evidence" pointed to oligopoly as the "ruling market form in the modern economy." The economist Joe Bain, who authored the leading antitrust text of the postwar period, noted the "ubiquitous category of oligopolistic industries" in "our predominately oligopolistic economy." Reflecting such views, and tapping an agrarian creed dominant from the nineteenth century until at least his son's bid for the presidency in 1984, the Reverend Theodore Mondale believed "the greatest danger confronting capitalism is the ever increasing concentration of wealth in the hands of a few. The concentration of wealth gives an undue power to a well organized group for economical and political control and for further concentration of wealth and complete control." South Dakota rancher Homer Ayres agreed with his grandfather, and with Reverend Mondale, that "it was the monopolies, or the 'money power' that was the real enemy of the common people."[1]

A large body of theoretical literature and political thought addresses the idea of "monopoly capitalism," viewing the continuing concentration and conglomeration of economic power as an inevitable characteristic of "late capitalism." Paul Sweezy, who coauthored the book *Monopoly Capital* in the 1960s, believed he understood the cause of postwar American capitalism's problems: "The reason is that the process of monopolization—what Marx called the concentration and centralization of capital—is a continuing one which has characterized the history of capitalism throughout the present century and is still operating." The economist Joseph Schumpeter had already conceded in the 1940s that "Marx's vision was right."[2]

Concentrated, uncompetitive, unworkable markets signaled "capitalism in decay," a stage when business turns to the state to revive a flagging economic system and for protection against revolution, the prelude to fascism. Noting fears that the "concentration of industrial power may lead to the police state," the head of the Antitrust Division of the Department of Justice once asked: "Can anyone doubt that the prewar experience of Germany, Japan, and Italy have proven the wisdom of the nation's concern over concentration of economic power?" Writing in the wake of German fascism and Japanese imperialism, the Chicago economist Henry Simons argued that private monopolies were dangerous because they promoted "an accumulation of government regulation which yields, in many industries, all the afflictions of socialization and none of its benefits; an enterprise economy paralyzed by political control; the moral disintegration of representative government in the endless contest of innumerable pressure groups for special favors; and dictatorship." Even before the war Harold Ickes warned Americans in an NBC radio address about a "Big Business Fascist America."[3]

Similar fears among farmers and their champions shape the stories of postwar rural America, stories involving talk of the rural "rot belt," "rural Buchenwalds," "Appalachia in the Heartland," and hopeless "wastelands"—denoting decay, fascism, backwardness, and death—all stemming from the encroachment of monopoly capitalism, in which "agribusiness" played an important role.[4] I propose here to analyze the nature of the monopoly problem as it affects the farmers of the American grain belt after the end of the Korean War. The area of inquiry is part of what Lincoln called "the great interior region." More specifically, it is what Frederick Jackson Turner called the Old Northwest Territory states' "trans-Mississippi sisters of the Louisiana purchase—Missouri, Iowa, Minnesota, Kansas, Nebraska, North Dakota . . . South Dakota," and I include Wisconsin. This region constitutes the "grain belt," where, with a few other states, roughly 80 percent of the nation's soybeans, corn, and hogs are produced, in addition to a large portion of the nation's wheat and cattle, all commodities to which I pay particular attention. The study begins about the time the Korean War ended, but covers previous events in places. The 1953 Korean armistice ended the "longest period of sustained prosperity in American agricultural history," a period that started when the Nazis invaded France. Global crop production also reached near-record levels in 1953, coupling with the falling demand for food after the end of the war to put tremendous downward pressure on prices for the farmers' products, creating the "farm problem" that would greet President Eisenhower and his successors. The study ends around 1980, when Ronald Reagan's electoral revolution completely changed the center of gravity in American politics, signaling the "fall of the New Deal order." The early 1980s also witnessed a particularly acute agricultural depression that triggered the creation of 100 to 150 new farm organizations and transformed the landscape of farm politics, distinguishing it from the 1953–80 period. Moreover, the research materials for the post-1980 period are much more scarce. Even though historical records after 1980 are more sketchy and limit research in this time period, the public policy issues surrounding the monopoly problem persist to this day.[5] Given the dramatic increase in concentration levels in agricultural processing within recent years, the monopoly problem is as prominent as ever. In April 1999 I attended a meeting of farmers and the head of the Department of Justice's Antitrust Division in South St. Paul, Minnesota. Senator Tom Harkin (D, IA) commented that it was the first time in his nearly three decades in politics that the head of the Antitrust Division had met with farmers, a measure of the continuing concern among farmers about the monopoly problem.

THE PROBLEM

In the latter half of the nineteenth century American farmers grew concerned about increasingly erratic agricultural prices and growing economic concentration in the industrial sector. After a trip to the country to meet with farmers in the late 1860s, the founder of the Grange returned to Washington and reported, "They all want some plan of work to oppose the infernal monopolies." Such sentiments inspired widespread attempts at farmer cooperative formation to stabilize prices and match industrial market power and helped secure passage of the Sherman Antitrust Act to combat industrial concentration. Into the twentieth century farmers continued their dual strategy of advocating organized, cooperative commodity marketing for themselves while promoting competition in the industrial sector in an effort to reach economic "parity" with the "industrial and commercial domination" of the corporate economy.[6] Despite greater government intervention during the Great Depression, monopoly fears persisted as in-

4 THE PROBLEM

dustry continued to concentrate and efforts to secure additional political relief or achieve self-organization continued to unravel. Farmer dependence on a few input producers and the threat of "agribusiness" integrating backward into agricultural production compounded the postwar problem.

The two-part problem involves the desirability and workability of competition. The first part includes the attempt of farmers to stabilize a fluctuating market economy that kept them in constant fear of losing their farms.[7] Their efforts involved some sort of monopolistic control over the market to reduce output and raise prices, whether it be through government-granted monopolies by means of marketing quotas, state-planned production, the private organization of cooperative marketing, farmers collectively bargaining for contracts with agricultural processors, or other means of relieving the "inequality in bargaining power that exists in farm product markets." The second part of the problem involves the workability of competition in sectors of the economy closely linked to farming. Foremost in this category are food processing, meatpacking, and exporting, which have all been accused of tending toward oligopoly and monopoly, increasing the "disparity of farmer bargaining power." Also significant to this half of the problem is the potential for outright takeover of agricultural production by these sectors' attempts to integrate vertically or by the growth of nonfarm corporations involved in agriculture.[8]

The problem originates in the nineteenth century. After the Civil War, railroads brought a greater commercialization of farming, dwarfing any previous market participation, and transformed "agriculture in the United States from the family-oriented, self-sufficient farm to commercial agriculture serving distant consumers."[9] Further compounding the market orientation was a growing advocacy of business values in farm journals and farm newspapers, enhanced usage of scientific and technical innovations, and productivity improvements, all of which contributed to the cultivation of large tracts of new land, more than the increase for the previous 260 years.[10]

The ascendant managerial and technocratic values that shaped farming at this time also affected the industrial sector of the American economy. These values, coupled with new technology, materials, and marketing techniques, however, also contributed to chronic price instability, "social chaos," and "destructive competition."[11] In response, trust formation and monopolization became common. In the 1880s, for example, the railroads came to be "dominated by a relatively few huge railroad systems, each operating several thousand miles of track." Other monopolistic industries specifically affecting farming included meatpacking, barbed wire, cottonseed oil, linseed oil, sugar, whiskey, biscuits, and the cattle trust, to name a few. Organized agrarian protest in the United States started as a response to this monopoly problem and with its promotion of railroad nationalization and government regulation, and its organization of cooperative marketing to temper the market represented the first serious challenge to the orthodoxy of classical economic liberalism in the United States.[12] Out of this heritage emerged agriculture's contemporary monopoly problem, whipsawed, as many farmers saw it, by a chronic inability to organize economically and a dependence on a widening yet concentrating industrial sector.

From the beginning of the American republic, what Joyce Appleby calls "ancillary trades," the "millers, teamsters, ship and wagon builders, bakers, coopers, and grain merchants [which] sprang up to process, transport, and sell American grains," were a concern to farmers. Historians have typically echoed farmers' concerns and talk of "institutional leviathans—the corporations and trusts—with centralized control over the economic lives of farmers" and "corporate monopoly" setting prices by "administrative edict." The classic comparison describes farmers marketing in an extremely competitive and disorganized market, one economists hold up as the example of "perfect competition," while buying in concentrated markets and selling in oligopsonistic markets. Whereas farm prices dropped 63 percent and production only 6 percent from 1929 to 1933, for example, the price of farm implements fell only 6 percent while production dropped 80 percent. Contemporaries believed these price statistics proved that the "'modern industrial organization' had 'destroyed the free market' and lodged the making of industrial policy in the hands of a few private individuals."[13]

Corporate mergers and acquisitions after World War II fueled these long-standing fears of industrial concentration and prompted continued talk of substituting "public regulation [for] the corporate giants' private planning" or of divesting large firms like General Motors and U.S. Steel.[14] For farmers the specific concern was the food-processing industry, a key component of "agribusiness," the postwar term describing the "sum total of all operations involved in the production and distribution of food."[15] Already described as tending toward "monopoly and oligopoly" in the 1930s, fifteen years of mergers between 1950 and 1965, according to some government reports, left 80 percent of value-added food products in oligopolistic industries. Many believed that under these conditions farmers would never get a fair price and that most of the profits from food would be enjoyed by the processing industry.[16] Similar logic was applied to the aspect of agribusiness handling grain exports. Although the industry involved thirty-some firms in the 1920s, six companies exported 96 percent of wheat, 95 percent of corn, 90 percent of oats, and 80 percent of sorghum in the 1970s. The massive postwar rural to urban migration—the largest "the world has ever known" according to President Kennedy's secretary of agriculture—also coincided with a growth in corporate farms from 1960 to 1968 that matched the total for all previous time periods, causing one U.S. senator to predict rural America's "headlong descent into a state of corporate feudalism."[17]

The opposition to corporate farming and monopoly was part of a much broader dilemma. As Ellis Hawley has argued, the problem involved larger "questions of power . . . the development, in particular, of private concentrations of economic power and . . . the implications of this development for a democratic society." For some, the postwar plight of the American farmer epitomized this problem; the classic exemplar of American republicanism, the yeoman farmer, faced concentrated, collusive, and uncompetitive markets and corporate buy-out efforts. The struggle against the various manifestations of the monopoly problem, so it was argued, became the last hope of preserving any semblance of a competitive economy or the nation's democratic principles.[18] The end of the independent farmer also meant the deterioration of agriculturally dependent rural areas and small towns—the "backbone of America," where the "American tradition of democracy was formed"—while those displaced were forced to "rot on the welfare rolls in urban slums." As David Lynch noted about the Great Depression, "landless men, great armies of 'Joads,' constitute a festering sore on the social and political body and contribute a poor foundation to a political democracy."[19] Those "tractored out" in the postwar period were no different and no less detrimental to democracy. Their plight prompted dark humor of rodeo as a "refugee camp for cowboy wannabees dislocated when farming and ranching became agribusiness" and more sober descriptions of "technoserfdom" and the "rise of America's rural ghetto."[20]

Compounding the damage to agrarian democracy and small-town values, opponents of corporate farming argued, were the dangers and disproportionality of large-scale industrial production techniques. The logic resonated, given the heightened environmental awareness of the postwar period and the long-standing "dominant image of an undefiled, green republic, a quiet land of forests, villages, and farms dedicated to the pursuit of happiness."[21] Consider a literary treatment:

> And then one day we were running southwest from Dodge City after a very bad morning in west Kansas, a morning starting in Scott City and soon lost in the depressing agribusiness haze and swamp, in the humiliating funk of that rampant out-of-proportion agriculture and its miles of chemical-soaked fields and giant rococo machinery, its outrageous absentee-landlord devastation of the Arkansas River bottom east from Pierceville, wheat jammed in to the very banks, leaving not a single tree for 20 miles along one the continent's major rivers. It is a fouled mess and monotone, an ignoble extreme of slash and burn.[22]

Some farmers wanted to leave the arduous life of the farm and seek work in the cities, but many did not. A year after World War II a Gallup Poll asked about the "chief faults of city people" and found that farmers thought city people "feel too superior," were selfish, and "live[d] too hurried a life." A 1965 poll by the Minneapolis Tribune indicated that 94 percent of farmers preferred life on a farm to life in a city home.[23] A Minnesota farm wife wrote to her governor in 1964: "Some say if you aren't satisfied get off the farm do something else. Sure we can do that but does that solve the farm problem? We love the farm and have no desire to do anything else. There are getting to be fewer and fewer that want to stay on the farm and the way its going those of us who are willing to withstand almost anything because we love the farm will not be able to stick it out." Agonizing over the possibility of leaving the farm, a North Dakota farmer wrote to his senator: "I am a farmer out here in this Confused Country. I don't know what to do anymore. One day our hopes are raised & the next day the bottom drops out of everything." Another North Dakota farm wife sent off a letter saying, "Our farm prices are so out of line with the cost of production that most farmers are either going to have quit farming altogether, or find some income else where."[24]

Many farmers did leave the land, contributing to large-scale changes in the grain belt political economy (see table 1). Farm counties in eastern South Dakota, for one example out of many, saw large-scale outmigrations. Sandborn County, in the area of the state that produced farmer advocates such as George McGovern and Hubert Humphrey, lost 27 percent of its population in the 1950s, leaving less than forty-seven hundred people in the county. As Humphrey was running against McGovern in Democratic presidential primaries in the 1970s, he linked rural depopulation to

TABLE 1. CHANGES IN FARM STRUCTURE IN THE GRAIN BELT, 1950–1992

	Number of Farms, 1950	Number of Farms, 1992	Average Size of Farms, 1950 (acres)	Average Size of Farms, 1992 (acres)
North Dakota	65,302	31,123	535	1,267
South Dakota	66,331	34,057	575	1,316
Nebraska	107,174	52,923	442	839
Kansas	131,372	63,278	370	738
Minnesota	179,119	75,079	184	342
Iowa	203,155	96,543	168	325
Missouri	229,958	98,082	153	291
Wisconsin	168,582	67,959	138	228

Source: Bureau of the Census, *General Report, United States Census of Agriculture, 1950*, vol. 2 (Washington DC: GPO, 1952), 824–25; Bureau of the Census, *Census of Agriculture, 1992*, vol. 1, *United States Summary and State Data*, Geographic Area Series, part 51 (Washington DC: GPO, 1994), 170–75.

the wider social convulsions of the 1960s and 1970s, arguing that rural outmigration "has festered the social and economic sores which have erupted around us." McGovern's South Dakota colleague in the Senate, James Abourezk (D), reflected on the plight of places like Sandborn County: "[Rural America] is the place left behind. It is dying on the vine, a victim of strangulation by social, political, and economic neglect. That neglect is propelling us, gradually but powerfully, toward national human disaster. There are now on the landscape thousands of dead small towns.... [The] small towns on the land in the middle of my State are literally drying up and blowing away."[25]

The difficulties of making a living on the farm prompted many grain belt states to launch "economic development" programs to diversify their largely agrarian economies and create new employment. Generally, states such as South Dakota, to continue the example, benefited from the construction of the interstate highway system in the 1950s and the building of dam systems on the Missouri River and, coupled with their economic development programs, were able to increase the number of people employed in manufacturing from about ten thousand in 1947 to twenty-three thousand in 1977 and to thirty-five thousand in 1992. States became "entrepreneurial" in courting industry, and governors became consumed with the effort of courting business, prompting commentary about the "political economy of seduction," a potential "race to the bottom," and an "economic war among the states" seeking industry.[26]

Despite some successes, historians have doubts. William Pratt believes that "neither Nebraska, Iowa, or the Dakotas have truly succeeded in their economic development endeavors." Mark Friedberger concluded that "the dog-eat-dog competition in the rural Midwest underlined the huge amount of wasted effort which smaller communities gave to the problem of trying to provide jobs for its citizens." And some wondered about the benefits of some development efforts, such as the movement of large-scale meatpackers into small towns and greater reliance on industries such as gambling. Whatever the level of success, the anxiety of the North Dakota farm wife about being forced to find work off the farm could not be quelled. As one Iowa farmer noted, the limited amount of work produced a simple social equation: "leaving the farm . . . means leaving the county." When the Kennedy speechwriter and native Nebraskan Theodore Sorensen traveled to his home state in the 1960s, he reflected on the grim prospects for rural America, proclaiming that Nebraska had become "a place to come from or a place to die," underscoring the urgency of the rural states' economic development initiatives. An outgrowth, according to some, of farmers' inability to solve the monopoly problem, rural economic development became another method of engaging the problem. The longtime agricultural economist Harold Breimyer once asked, "Why then should the rural community be developed?" and answered his own question: "It should be developed as a countermeasure to the relentless national trend to envelop everyone and everything into a giant, faceless, mechanistic, conglomerate-corporation urbanized bureaucracy."[27]

RETHINKING THE PROBLEM

A recent interpretation of business in the 1920s and 1930s argues that scholars have tended to "confuse organizational rhetoric with organizational strength" and underestimate the "pervasive competitive disorganization" and "chaos of the market," highlighting the stumblings of business-sponsored trade associations lacking "powers of enforcement or persuasion" and casting serious doubt on the ability of business to control any prices.[28] New evidence also exists of competition in the export trade and of declining single-firm concentration ratios in industries such as agricultural implements, flour milling, meatpacking, and sugar refining. The standards of measure have also changed. The corporate economy used to be compared to an ideal economy of pre-1860 America. Economic historians, after studying the degree of competition in local markets, now argue that "many industries in the 19th century were probably as concentrated as, or more concentrated than, in the 20th century—within the relevant market boundaries."[29] Most importantly, the theoretical foundation of past policy making and historical interpretation has been seriously challenged by a new understanding of monopoly. Led by Robert Bork and Richard Posner, the "Chicago school" of antitrust undermines the notion that markets defined by four or five major firms, such as many food processing markets, are uncompetitive and thus challenges the long-standing New Deal–Warren Court antitrust theories frequently cited by agricultural historians. The older theories assume that firms in concentrated markets will not cut prices or increase output because they fear triggering a ruinous price war. Instead, the new scholarship argues that market structures tend to reflect efficient and competitive outcomes and highlight the impossibility of the cooperation needed to maintain cartel arrangements that reap monopoly profits.[30]

Such developments are considered in my analysis of the postwar monopoly problem. Instead of assuming the accuracy of "oligopolistic interdependence" theories and citing the four-firm concentration ratio as an indicator of anticompetitive behavior, which amounts to conviction by correlation, I take into account indicators of competition such as stability of market shares, substitutability, demand elasticity, foreign competition, and barriers to entry, applying archival evidence where possible. Instead of the "playing of games on the blackboard and computer," I pay attention to "the experiments of history," assuaging dead economists "gazing down from Valhalla" who would think that "economics with the history left out" was "bizarre," and answering a question that should be "ask[ed] of every dissertation . . . so what? What have you taught me about the actual economic world?"[31] Assuaging Americans who pay the university bills is also important; a republic needs some history and economic analysis "comprehensible to its own citizens."[32]

The empirical approach also considers the oligopoly theories prominent in much of the Chicago scholarship.[33] Willard Mueller, for example, an economist at the Federal Trade Commission in the 1960s, assumed a "perfect correlation between market concentration and market power, and consequently, profit rates."[34] He believed that the degrees of market concentration set forth in Joe Bain's *Industrial Organization* (1959) represented the "current consensus of scholars of the general area of industrial organization and performance."[35] Robert Lanzillotti thought "both economic theory and empirical studies strongly support[ed] the view that structural characteristics of industries [were] the significant determinants of market conduct and performance."[36] Combined with the New Deal–

era conception of how firms in concentrated markets behave, the market structure analysis became the "best single, generally available, measure for evaluating the importance of monopoly."[37] The use of the theory can be seen in President Johnson's White House Task Force on Antitrust Policy recommendation that Congress pass the Concentrated Industries Act. The legislation directed the attorney general to initiate legal proceedings against all "oligopoly firms" (a firm with 15 percent of a market in which the four-firm concentration ratio was 70 percent) in order to reduce them to 12 percent or less of the market.[38] Using the same logic, the Supreme Court decided the 1966 Von's Grocery case, in which the contested merger resulted in a firm with 1.4 percent of the grocery stores and 7.5 percent of the grocery sales in Los Angeles. The court declared it illegal.[39]

Structure can be a poor indicator of competition, however. The Big Four meatpackers, condemned as the "greatest trust in the world" by the Populists, experienced a degree of competition after World War II that shattered the previous market structure. In the late 1970s, when the FTC debated launching "The Bride of TNEC" (the Temporary National Economic Committee studied corporate concentration in the 1930s) part of the reason was the confusion over the "antitrust dilemma" resulting from the conflicting studies over whether "structuralist" antitrust prevented or promoted inefficiencies.[40] It is extremely difficult to predict how markets will develop or function, hence the "long and frustrating search in the dark" for a plausible "theory of oligopoly" and the need for empirical research, and less theorizing, into the historical development of markets.[41] While maintaining hope of finding some helpful theories, a leading industrial-organization economist concedes that "virtually anything can happen."[42] And hence the sophisticated political and theoretical prognostications of some academics about the "evolutionary path" toward monopoly capitalism are wrong. It offers another reason, as Richard Rorty argues, for intellectuals "to rid themselves of the idea that they know, or ought to know, something about deep, underlying forces—forces that determine the fates of human communities."[43] In the case of the monopoly problem, the indeterminacy of certain economic analyses creates more room for normative political judgments by citizen-legislators in areas such as antitrust law.

The problems of the structural school also haunt the Chicago school. Although structure cannot be definitively linked to anticompetitive behavior, it does not follow that oligopolies are always competitive. In spite of economic arguments about the difficulties of cooperation and the tendency to cheat among cartel members, collusion in oligopolistic settings can work.

In the most famous case, OPEC, oil exporters are often able to reap monopoly profits by colluding. Closer to farmers, the ready-to-eat cereal (RTE) market in the 1960s and 1970s seemed to maintain a system of price leadership that fostered collusive behavior. Kellogg led twelve out of fifteen cereal price increases between 1965 and 1970; General Mills followed the price increase nine times, and Post followed ten times. Courts have noted the continuation of price leadership in the industry in more recent years. In another market important to farmers, lysine, Archer Daniels Midland was found guilty of fixing prices during the 1990s in a three-year international conspiracy with four Asian companies. When ADM and Ajinomoto executives met in the infamous smoke-filled room, competitive Chicago assumptions did not prevail: "So the question is how do we share th[e] growth [in the lysine market]? What would you be willing to accept and what would we be willing to accept?"[44]

The multiple outcomes produced by economic theorizing offers very little concrete guidance to judicial rule makers. In a pioneering article that contributed heavily to the early interpretations of Celler-Kefauver Antimerger Act of 1950, Derek Bok highlighted the problem of relying on economic theory. Given the "aura of complicated uncertainty" surrounding the competitive effects of a merger, Bok believed that reliance on economic analysis would cause "confusion rather than enlightenment." Bok believed that by attempting to incorporate economic "expertness we may only end in extravagance."[45] Economic assumptions about the behavior of actors in particular contexts, as the Supreme Court has more recently conceded, must be accepted "on faith."[46] The embrace of economics, according to the current chairman of the FTC, requires the incorporation of "large doses of hunch, faith, and intuition."[47]

With the embrace of economics, according to Frederick Rowe, the former chair of the American Bar Association's Antitrust Section, antitrust law "bound itself to a delusion." Rowe blames the New Dealers who embraced the economics of the 1930s and then prominent theories linking economic concentration to inefficiency and corporate sloth. A core component of deconcentration efforts in the 1940s involved "antitrust law's assimilation of economics." But over time the economics changed. The coming of the Chicago school and its emphasis on economic efficiency produced different results in antitrust cases. Economics, according to Rowe, fulfilled the "Faustian pact of the forties" and "the servant bec[ame] the master, first abetting, then usurping, antitrust law."[48] This usurpation has left the antitrust laws, in the words of Justice William Douglas, "mere husks of what they were intended to be."[49]

Breaking the theoretical stalemate requires the consideration of additional factors. Thus the other component of this work, farmer organization, what John Kenneth Galbraith called countervailing power. In addition to the frequently mentioned issue of high concentration, I want to take into account the relative organizational strength of farmers. George Stigler mentions the effect of the number of buyers on the ability of sellers to collude and conspire, but he doesn't look at how the organization of oligopolists' suppliers affects the ability to collude and conspire.[50] Another analysis lumps the question of supplier organization into the "other sources of market power" category. The author mentions that "firms may have market power conferred upon them by weakness, usually temporary, of those with whom they trade," but the idea only warrants two sentences in a dismissive paragraph. In another part of the article the author (in parentheses) mentions the problem of monopsony power and uses the example of canning companies purchasing vegetables, but he mentions the issue only to make another point.[51]

The idea of promoting countervailing power as public policy developed in a haphazard way in the 1930s and received its fullest distillation in Galbraith's 1952 book *American Capitalism: The Concept of Countervailing Power*, which noted that "the notion that there might be another regulatory mechanism in the economy [countervailing power] has been completely excluded from economic thought."[52] How vertical sectors of equal power, what economists call bilateral monopoly, would work themselves out is unclear. The economist Willard Mueller wrote to a colleague in the late 1970s that "it is now generally accepted among industrial organization economists that a competitive outcome is only one of many possible solutions to bilateral monopoly.... Simply put, the result is indeterminate." Others doubted that, even if farmers were well organized, they could increase their incomes significantly by bargaining away the modest profit margins in the food processing industries.[53] Since it is very hard to know how bargaining would change things for farmers we must go to the record and find the available evidence, instead of relying completely on the blackboard.

Organizing farmers into a countervailing force was complicated. Unlike industrial workers, the first example Galbraith cites, the economic interests of farmers are more varied. Some farmers raise corn for sale and consequently want the price to be high, while others feed it to hogs and cattle and consequently want the price to be low. Some farmers consider their farms firms, or "agribusinesses," and themselves both management and labor, "neither exploiter or exploited," blurring the "delineation into farm and nonfarm sectors," a blurring represented by the title of the 1957 book *Farmer in a Business Suit*.[54] The blurring existed since the beginning of farm protests over the monopoly problem, as indicated by the Great Commoner's most famous speech: "We say to you that you have made the definition of a business man too limited in its application . . . the farmer who goes forth in the morning and toils all day . . . is as much a business man as the . . . few financial magnates who, in a back room, corner the money of the world. We come to speak for this broader class of business men."[55] The ambiguity over the distinction between farm and agribusiness can be seen in Senator Robert Dole (R, KS), who was called both "Iowa's Third Senator" and the "Senator from ADM," the former for passing a farm bill during the 1980s recession and the latter for supporting ethanol subsidies, both seen as beneficial to agriculture. But farms, unlike many firms, must make planting decisions long before demand is known, cannot change crop production or herds easily to meet changing demand, and must deal with the vagaries of weather, disease, insects, and animal biology. The complexities are reflected in the congressional "farm bloc," which ruptured in the postwar period over commodity differences, region, income, and partisanship, when farmers were accused of "speaking with too many voices." On top of these divisions, farmers are very independent, often unwilling to accept the controls involved in economic organization.[56]

Despite these and other problems, farmers in the postwar period were able, to an extent, to organize and improve their marketing techniques. In the grain belt, the National Farmers Organization successfully bargained for higher prices, and cooperatives such as Farmland, Grain Terminal Association, and Far-Mar-Co also made great gains in the postwar years. In the 1970s 26 percent of all commodities were cooperatively marketed, and six farmer-owned cooperatives enjoyed Fortune 500 status. The successful organization of markets by various cooperatives actually prompted the FTC—an institution that cut its teeth investigating meatpackers after farmer complaints—to launch a study of over sixty farmer cooperatives during the 1970s.[57]

The idea of countervailing power adds another dimension to antitrust. If public policy promotes the building of countervailing power in certain sectors, then it should limit the concentration and conglomeration of economic power in the sector being countered. This could mean applying a tighter merger policy to that sector, despite potential losses in economic efficiency. Chicago economics, however, which finds most mergers efficient, has undergirded much of American antitrust policy since the 1970s, obviating the political uses of antitrust as a tool for building countervailing power. "Economy" has trumped "political" in the political economy of antitrust.

Antitrust laws, any laws, in a republic are necessarily political. The New Dealer Robert Jackson's antitrusting was driven by his childhood in upstate New York, a "socially classless" society, "truly and deeply democratic, democratic in an economic and social as well as political sense . . . the nearest Paradise most of us ever know." His objection to the arrival of chain stores was "not economic so much as it [was] a social objection. The question is whether we have paid too much for these economies," payment in the form of "old friends who ran the several stores and performed their important functions in the community. . . . We do not like to have any one man or corporation own the town." As World War II was ending, Judge Learned Hand agreed, concluding in the famous Alcoa case that the antitrust laws were designed "to perpetuate and preserve, for its own sake and in spite of its possible cost, an organization of industry in small units."[58] A Truman administration official saw Republican efforts to limit funding for the FTC as "an excellent political issue," and Richard Nixon, the consummate politician and also the son of a small grocer, understood what the official meant. In his first year as president, Nixon supported anticonglomerate merger suits against ITT in a meeting with the Council of Economic Advisors: "This is a tremendously potent political problem which doesn't mean we don't tackle it. Does it mean that Mom and Pop stores are on the way out—and supermarkets are all we have? There is a sociological problem here. We may be helping consumers, but we don't help the character of our people. This is an old fashioned attitude . . . I know—but I would rather deal with an entrepreneur than a pipsqueak manager of a big store."[59]

Groups such as the National Farmers Union worried about how the new Nixon administration would approach antitrust, fearing that the New Deal, structuralist version would whither in the face of growing criticism and doubts. Alan Greenspan, who advised Nixon, authored a chapter in Ayn Rand's *Capitalism: The Unknown Ideal* calling for "the entire antitrust system [to] be opened up for review," characterizing "the entire structure of antitrust statutes . . . [as] a jumble of economic irrationality and ignorance. It is the product: (a) of a gross misinterpretation of history, and

(b) of rather naive, and certainly unrealistic, economic theories."[60] The president of the South Dakota Farmers Union feared that Nixon would forget the sociological side of antitrust and instead view it as solely a matter of economics, as did Greenspan and others, and he told the Senate why: "You can drive anywhere in the rural areas and see the results of our failure to weigh social consequences in determining our economic objectives: the weathered, abandoned farmhouse, a curtain flapping through a broken window; the soaped-up plate glass of the store front with the 'closed' sign taped to the door; the weeds standing tall around the vacant service station, and the growing ratio of older people on our main streets in areas like South Dakota." Thomas Jefferson would have understood, for as A. Whitney Griswold has written, "agriculture, to him, was not primarily a source of wealth but of human virtues and traits most congenial to popular self-government. It had a sociological rather than an economic value."[61]

The economist Donald Dewey concedes that the "view that Congress intended the primacy of [economic efficiency] when it passed the Sherman Act in 1890 is historically suspect and can only be supported, if at all, by a highly selective use of evidence." We are, after all, a "country that is dedicated to the preservation of bicameral state legislatures (even in Rhode Island), the Electoral College, independent sewer districts, juries in civil cases, township governments, grand juries, and popular election of coroners" for a reason, because they "disperse power and ... decision-making." When urging attention to the "question of tradeoffs" in antitrust, Dewey mentions the costs of antitrust's anticompetitive effects versus the benefits of its deterrence to price-fixing. Moving beyond the economic tradeoffs, Dewey also considers "honorable values" such as "decentralization of decision-making, the dispersion of power, and a higher standard of business ethics." "There is no reason why the Law's reasonable man should not conclude that antitrust, with all its presumptive inefficiencies, is not worth the cost."[62] Sometimes the largest variety of shirts at prices approaching cost is not the most important priority. Many people shop locally, discounting the "welfare" gains of going to the big city, so the downtown of their community will survive, buying from Bob the hardware guy who sits next to you in church instead of Sam Walton. The Chicagoan Deirdre McCloskey admitted after a recent trip to Europe that "economic tourism doesn't always come to Chicago-School conclusions."

> European cities are pretty, much prettier than American or Australian cities. Why? Zoning laws and building codes that Americans and Australians would regard as fascistic. In most European cities, large and

small, the old downtown is still where people shop, and not because they are quaintly attached to The Old Way. The town councils do not allow strip malls to develop on the outskirts. So the outskirts are charming country scenes and the downtowns are charming urban relics. And I say—I bite my lip when I say it, but I say it still—there's *something* to be said for it. If you could see Gouda and Delft you'd wonder whether we libertarians should fight to the death for the Coralville strip in Iowa City.[63]

Others have urged political conservatives to abandon their hostility toward government antitrust policies, viewing them as a way of avoiding more technical and intrusive kinds of government planning, preserving the citizen's faith in market capitalism, and respecting "society's abiding concern for the diffusion of private power and maximum opportunity for individual enterprise." When Publius tried to explain how our experiment in self-government might work, welfare maximization and efficiency gains took a back seat to spreading, diluting, and dividing interests, parties, and factions, incubators of "instability, injustice, and confusion," the "mortal diseases under which popular governments have everywhere perished."[64] Decentralization as an antitrust goal is another way to combat these diseases.

A history of farmers and the monopoly problem, like histories written in the early nineteenth century, is "charged with the special task of inducing in men an awareness that their present condition was always in part a product of specifically human choices, which could therefore be changed or altered by further human action in precisely that degree."[65] It might, as Richard Hofstadter once said, have "something to say that might help us."[66] It is an important goal, especially when we've reached the absurd moment when a president of the Organization of American Historians has to plead with his colleagues to acknowledge "the pertinence of political history" and another historian begs his colleagues to concede that "political economy still matters," both hoping to redress what Elizabeth Fox-Genovese and Eugene Genovese call the "massive attempts by social historians to deflect attention to the bedrooms, bathrooms, and kitchens of each one's favorite victims." Those responding to such criticism argue that "in most university towns nowadays people . . . [are] concerned with women now, with lifestyles, and why everyone's so unhappy." Graduate students at Princeton organized a conference for October 1997 entitled "Casualties of History: Losers, the Lost, and the Problem of Defeat."[67] Perhaps this focus of academic history is linked to our democratic "crisis," our lack of attention to such public matters as the "palpable despair and cynicism and violence" in the republic and the "contemporary incapacity of American politics," what Christopher Lasch, just before his death, called "the democratic malaise." If the "end of history" is here, one of democratic capitalism, we had best get on with the conversation about which variety we want, the "search for a half-way house between Cobden and Lenin," as Beveridge said, and quit stewing about why we are so "unhappy."[68]

2

THE CORPORATE FARMING DEBATE

The man on 80 acres or 160 acres of land may be an outstanding citizen, a good neighbor. He is not small. Character, prudence, honesty, energy, and patriotism has been typical of the citizens on family type farms throughout our history. The citizens of our nation will not now stand by and see this type of citizenship replaced by Syndicated Corporation owned farms. DAN TURNER, FORMER GOVERNOR OF IOWA

Once the corporate landlords have all the land . . . then the era of peasant agriculture will have arrived. CHARLES WALTERS JR., *ANGRY TESTAMENT*

After World War II, when Ben Hogan's financial troubles worsened and his line of credit dried up, he borrowed money from the Nowell-Safebuy, which had purchased his mortgage from the bank. The loan contract stipulated that the turkeys he raised with the borrowed money could make their way only to the Nowell-Safebuy processing plant, where the company could choose the ones it wanted and refuse to buy the undesirables. Ben hated the arrangement: "There ain't a dirt farmer got a pot to piss in, what with prices are this year. . . . A man either keeps raising turkeys or he don't get no loans. A man can't make it without loans. There's no telling how many farms gone under that way since the war. And the Nowell-Safebuy ends up with 'em all." When feed costs increased, the farmers around Nowell, South Dakota, could not make their loan payments and filed a lawsuit against Safebuy. The judge, ruling that the contract gave Safebuy "undue bargaining power to set prices on the products it buys and undue power to depress prices in a regional market it virtually controls," refused to "enforce unconscionable bargains." The company then decided on a new corporate strategy: actual ownership of the turkey farms, bypassing the family farmers completely and creating a larger, more efficient, integrated corporate farming operation.[1]

Stories like that of the Nowell-Safebuy, told in Douglas Unger's novel *Leaving the Land*, permeated rural America after World War II. Although corporate attempts to take advantage of scale in American farming are a three-hundred-year-old story that include the Puget's Sound Agricultural Company, the Wheat Farming Corporation of Topeka, Kansas, the corporate "cattle kingdoms" of the Great Plains, and the famous Bonanza Farms of the Red River Valley, these attempts resulted in "almost consistent failure" and kept corporate ownership of agricultural land at a minimum. The postwar years, however, marked a "major turning point in American agricultural history" as the overall scale of farming rapidly expanded and the number of corporations operating large farms increased.[2] Unlike previous periods when "no real threats from monopoly power in agriculture" existed, some began to fear that farms would become very large and controlled by a few corporations, surrendering control of the nation's food supply to corporate America. "You would end up with what they had in Poland," according to the manager of the Grain Terminal Association, "where a large number of great big, fat landlords owned the land, and it was worked by millions of peasants—complete feudalism."[3]

Fear of the coming of large-scale, corporate farming to the grain belt drew on postwar anxieties over the changes in American farming, disruptions in rural communities, and traditional political sentiments about outsider control. Many feared that corporate conglomerates would seize greater and greater tracts of farmland, drive off independent producers, control food from "seedling to supermarket," and ultimately collude to fix food prices at high levels, gouging consumers. The fears resulted in a series of studies and long public debates over the issue and prompted significant legislative activity, especially at the state level. In retrospect, however, some contemporary warnings about the "emotional appeals" of the anticorporate farming advocates and the "crisis atmosphere" they created now seem justified.[4] The changes that took place were influenced more by the consolidation of existing family farms than by any outside "corporate invasion." But the debate over corporate farming did underscore the noneconomic dimensions of the monopoly problem, and in the "emotional appeals" of the anticorporate farming advocates can be found a profound sense of regret about the changes in American farming.

The debate persists. In 1982 Nebraska passed Initiative 300, amending the state constitution to allow only family farm corporations to engage in farming. In 1983 Jim Hightower, the anticorporate farming activist, was elected agricultural commissioner in Texas. In 1988 South Dakota toughened its corporate farming law to prevent National Farms from establishing a large-scale livestock operation in the state. More recently, fourteen counties in Kansas put anticorporate farming laws on the ballot and they passed in twelve. More than a dozen family farm, religious, and environmental groups have been working to prevent large-scale hog operations from becoming the norm in Iowa.[5] And in the fall of 1998, the voters of South Dakota approved a constitutional amendment to restrict corporate ownership of farmland. The continuing strength of the anticorporate farming sentiment, coupled with the many changes in production agriculture after this study ends, requires that the issue continue to receive serious attention from policy makers.

THE FEARS OF CORPORATE FARMING

The debate over corporate farming was the most emotional dimension of agriculture's postwar "monopoly problem." Many believed the relationship between corporate farming and the depopulation of rural America to be causal, not coincidental. Leaders of the National Farmers Union (NFU) thought the "exodus from rural America" caused by low prices, attributable in part to the growth of large-scale farming and the outright entrance into agriculture of "giant, nonfarm corporations."[6] The National Farmers Organization (NFO) predicted that "corporate farms would wipe out commercial family farms by hundreds of thousands, and rural businesses and communities along with the farmers." The organization's president believed that the country was "losing free men" to a corporate agriculture that would soon control farmers' lives as the large growers in California controlled Mexican farmworkers. He viewed corporate involvement in farming as "Phase I of a corporate takeover of the food industry, which would involve acquiring or controlling all phases of production, processing and retailing."[7] The *Washington Post* noted fears of "20th century agricultural feudalism"; the president of the Agricultural History Society cited the fear of "a latter-day enclosure movement in the American countryside"; and a prominent agricultural economist feared "farming [would] be swallowed up" by corporate conglomerates "as nonchalantly as a pelican swallows a fish."[8]

The monopolistic implications of corporate farming led opponents to argue the importance of maintaining a competitive economy. The Agribusiness Accountability Project believed agricultural production "the last competitive sector of the food economy," and Senator Gaylord Nelson (D, WI) believed farmers the only suppliers in the economy "who [did] not have the opportunity to set what they believe[d] [was] a fair price for their commodities." The competition in the production sector, combined with the "lack of competition" in the overall economy, according to some, meant a thousand farmers a week were "squeezed out of business." Allowing corporate farming to grow would ensure that the "highly coordinated and integrated systems" of corporations would "take over" agriculture, ending the family farm tradition and guaranteeing that "food prices to consumers [would be] dictated by syndicates and not by competition."[9]

The arguments drew on the historical view that the proper form of the market economy was that of many, scattered, small-scale producers, a system perverted by the coming of big business in the late nineteenth century. The result was the concentrated wealth and power of the "moneyed men," the essence of the "monopoly problem" to some, in sectors like steel making and car building. The opponents of corporate farming feared that "agriculture [would] become—like steel, autos, and chemicals—an industry dominated by giant conglomerate corporations such as Tenneco." Protesting the establishment of a large-scale hog production facility in the early 1970s, the president of the NFO declared, "We are not going to allow a handful of corporate executives to control food production in the manner in which they now control oil, drugs, and autos."[10]

Given the events of the era, frequent comparisons were made to the competitive problems of the oil industry. Congressman David Obey (D, WI) asserted that "our troubles with a few oil giants should make us realize that if a relatively few farm operations take over, they can manipulate the supply and the price of food to meet their convenience—not the consumers." The central question, according to the Food Action Committee, was whether policy makers would "draw any lessons from the oil experience, or whether [they] will stand idly as that same way of doing business dominate[d] the food economy." Micki Nellis of *American Agricultural News* argued that "when the same companies which control the energy also control the food, they can bring any nation to its knees—including America." Senator James Abourezk of South Dakota agreed that "while monopoly control in food is not yet what it is in oil, all the symptoms are there." An oft-sighted bumper sticker on midwestern pickups: "If you think oil prices are high, wait till they own the farms."[11]

Such fears snowballed in an atmosphere of corporate distrust when the monopoly problem received greater public attention. In 1966, 55 percent of Americans expressed "a great deal of confidence" in business leaders, but by 1975 the percentage plummeted to 15 percent—during the nine-year period when the corporate farming debate was the most fierce. Big business even ranked lower in public confidence than the post-Watergate Executive Branch. *Harvard Business Review* reported that "public anger at corporations is beginning to well up at a frightening rate." Monopoly fears were so strong that Senate committees held hearings on the question of whether "the free, competitive market still exist[s] and still function[s] as a curb on excessive prices and as a stimulator of business and technical progress." Senators Nelson and Wayne Morse (D, OR) wondered if the market had not become a "myth carefully perpetuated by giant corporations to conceal the fact that they control the output of their industries, planning and regulating both prices and production in the privacy of their board rooms to serve their own ends." The National Federation of Independent Business feared the coming of "monopoly-socialism" and aided farmer advocates by endorsing legislation preventing corporate entry into farming. In the mid-1970s Senator Phil Hart (D, MI) led an effort to legislate government-imposed divestitures in concentrated sectors, and Congressman Mo Udall (D, AZ), who ran for president in 1976, called for a government commission to analyze certain concentrated sectors and to prescribe the proper remedies—deconcentration, regulation, subsidies to new competitors, whatever was needed.[12]

Enmity toward big business stemmed not only from questions about the monopoly problem. The anticorporate feelings of the 1960s and 1970s also grew out of the alleged involvement in prolonging the Vietnam War of companies such as Dow Chemical, a company similarly despised by those who objected to the environmental implications of greater farm pesticide use and the productivity changes that undermined farm prices. Student protesters in the 1960s, and many professors, seemed to adhere to a broad leftist critique of capitalism and the centrality of big business and the military—"the system." President Nixon exacerbated these views when he was caught with corporate-financed slush funds that paid for criminal activity; when his vice-president was caught taking envelopes in the White House basement; when his administration was caught working with the multinational corporation ITT to destabilize the socialist government of Chile's Salvador Allende and an antitrust case against the multinational was conveniently abandoned, all while ITT was often mentioned as a prominent corporate farmer. M. W. Thatcher, manager of the cooperative Grain Terminal Association, drew on these events and contributed to the cynicism about big business, declaring, "the Watergate experience may well prove to be the greatest eye-opener as to what actually happens in running our Government. Big Oil, Big Banking, Big Loopholes for the rich in our Income Tax Division demonstrate how Big Money, Big Political Contributions elect members of Congress and Presidents."[13]

Nixon's choice to be secretary of agriculture in 1971, Earl Butz, contributed further to the fears of corporate farming. Butz worked for the much-hated Secretary of Agriculture Ezra Taft Benson in the 1950s, advocated rolling back farm programs, encouraged larger farms, and served on the boards of agribusinesses such as Ralston-Purina and Stokely–Van Camp; he was barely confirmed by the Senate after a bitter debate. When anticorporate farming bills came before the Congress his department opposed them, earning him condemnation as an "apologist for corporate power" from the *New York Times*.[14]

The views of those farmers who felt they were being sold out to corporate agriculture seemed legitimized when it was revealed that some government officials were involved as corporate officers or consultants to agribusiness. Clifford Hardin, who was replaced by Butz as secretary of agriculture, took an executive position with Ralston-Purina when he left. Clarence Palmby, who was assistant secretary of agriculture when the Russian grain sales of 1972 were organized, afterward went to Continental Grain Company, one of the major companies involved in the transaction. Another assistant secretary, prior to taking his post, was a senior vice-president at Bank of America, which was involved in corporate farming investments. Virgil Wodika, before taking over the Food Bureau of the FDA, was a paid consultant to Ralston-Purina, Libby, McNeill & Libby, and Hunt Foods.[15]

Thus Congressman Jim Abourezk could tell the statewide South Dakota Farmers Union picnic in 1972 that USDA officials were "retreads from the Benson era or recent recruits from the corporate boardroom"; when running for president that year, Senator McGovern asserted that Butz "was thoroughly committed to the gentlemen farmers in agribusiness, who couldn't tell a chicken coop from a chain store"; Senator Fred Harris (D, OK) could argue that "the government has continually sided with the giant agribusinesses, turning its back on the little man"; Senator Ted Kennedy (D, MA) warned the Iowa Farmers Union that "we may be forced to watch corporate agriculture spread its tentacles to every farm in the nation"; when running for president in 1976, Jimmy Carter could denounce the "sweetheart arrangement" between USDA, big grain firms, and agribusiness; the Agribusiness Accountability Project could label the agribusiness-USDA connection as "Agri-Government"; and a farm couple could tell Senator Hubert Humphrey (D, MN) that "the Department of Agriculture & Administration & Big business guns [were] out to get [them]."[16]

Many believed that corporate influence also dominated the land grant colleges that were responsible for research and extension services for agriculture. The farm activist Jim Hightower argued in 1972 that "[the land grant college complex]—composed of colleges of agriculture, agricultural experiment stations and state extension services—has put its tax dollars, its facilities, its manpower, its energies and its thoughts almost solely into efforts that have worked to the advantage and profit of large corporations involved in agriculture." Instead of providing leadership on such issues as the corporate farming debate, "the land grant community has ducked behind the corporate skirt, mumbling apologetic words like 'progress,' 'efficiency,' and 'inevitability.'" When they did conduct research, many farmers and farm groups would not believe the studies because of the colleges' corporate connections.[17]

The federal Small Business Administration also received criticism. Created in the 1950s to offer low-interest loans to start-up businesses, the SBA made loans to individuals hoping to start livestock confinement operations, creating a great deal of hostility from farmers already established in livestock production. Senator Gaylord Nelson, chairman of the Senate Small Business Committee, criticized SBA support for hog confinements: "The SBA's loan practice in this area is especially disturbing because hogs are known to many farmers as 'mortgage burners'—low capital ways for young farmers to get a foothold in farming. As it stands, the SBA's policies are helping to fund the very factories which are driving people out of farming." IBP, the Iowa-based meatpacker that started with an SBA loan, became the packer most feared, and hated, by farmers in the postwar period.[18]

The suspicions and fears and prognostications of a "corporate takeover" of production agriculture were consistently legitimized by grain belt politicians. The corporate farming issue became a staple of postwar Democratic politics beginning in the 1950s; George McGovern of South Dakota is a good example of its uses. When changes in the poultry industry in the 1950s increased production and efficiency, they tended to drive egg prices down, triggering angry letters from farm wives about the dwindling amount of "egg money." McGovern responded by criticizing the "huge corporate interests" and "vertical integration" in farming.[19] During his 1960 Senate race against Senator Karl Mundt, McGovern told a Democratic fund-raiser that "if Nixon is elected, with men like Mundt who support him . . . the family farm is doomed as an institution and corporate agriculture will sweep the country." When he ran for president in 1972 he advocated his bill to "prohibit giant non-farm conglomerates from taking over family farms," and the Democratic platform stated that the "family-type farm is threatened with extinction. American farming is passing to corporate control." Politicians such as Senators Gaylord Nelson (WI), Hubert Humphrey (MN), Walter Mondale (MN), James Abourezk (SD), Harold Hughes (IA), Frank Church (ID), and Fred Harris (OK) all invoked the issue in similar ways. They drew on the antimonopoly tradition of the New Deal, highlighted by President Roosevelt's famous antimonopoly campaign in 1938, and on political and ideological arguments prominent in the grain belt dating back to the Grange in the 1870s, agrarian populism in the 1890s, and the Non-Partisan League around the time of World War I.[20]

In addition to elected officials and the warnings of prominent farm organizations such as the Farmer's Union and the NFO, religious leaders (especially Catholics), environmentalists, assorted writers, and network television were involved in the corporate farming debate. The rural-life director of the Catholic Church in South Dakota told the state legislature that "the family and ownership of land is the natural God-given way of human living and whenever the church or the state or powerful influential people forget that, and take over the ownership of God's land in a disproportionate manner, the economic, spiritual, and social balance of a nation is disturbed and evils of every kind result." The director of the Heartland Project of midwestern Catholic bishops asked that the "ideology of 'free enterprise'" be subsumed to "Christian, Jewish, and humanist perspectives that emphasize relationship, interdependence, and distributive justice, including fair compensation." In 1973 the Nebraska Catholic Conference advocated legislation to stop the "expansion of giant farm corporations" and called on Catholics in Nebraska to inform themselves on the issue, to celebrate Rural Life Sunday, and to support groups with similar views; internationally, the church adopted the view that "if certain landed estates impede the general prosperity because they are extensive, unused, or poorly used, or because they bring hardship to peoples or are detrimental to the interests of the country, the common good sometimes demands their expropriation."[21] In 1979 the president of the National Catholic Rural Life Conference reminded Catholics that "we are but sojourners and guests upon the Lord's land," promoted farming "as a way of life," and criticized "an agriculture characterized by industrialized farms with absentee owners [which] benefits a relatively privileged few and seriously weakens the nation's stability." The forty-four members of the Midwestern Roman Catholic Bishops also published the pamphlet *Strangers and Guests: Toward Community in the Heartland*, which proposed ending corporate acquisitions of farmland.[22] When the NFO, which had a disproportionately high number of Catholic members, started advocating collective bargaining for farmers as a strategy for preserving family farms and avoiding corporate agriculture, the church supported their efforts. Pope John XXIII even released a papal encyclical on agriculture promoting collective bargaining. The Red River Valley (Lutheran) Synod Convention, which hosted delegations from North Dakota, South Dakota, and Minnesota, also advocated the NFO's bargaining approach.[23]

The dangers and disproportionality of large-scale industrial production techniques associated with corporate farming also alarmed environmentalists. The fears took root with the heightened environmental awareness of the 1960s and 1970s, the perception of the family farm as a "countercultural response to everything decadent about industrial society," the hostility toward the "bigger the better" farming philosophy of Earl Butz and others, and the view that technological improvements doubled as "neighbor replacers." More fundamental was the view that big farms were "unnecessarily disruptive of the environment," used more pesticides and herbicides, generated more waste, and produced surpluses that depressed prices, hurting small farmers. Senator Nelson, an early advocate of environmental protections in the 1960s and 1970s, urged cooperation between environmental groups and small farmers. During debate over an anticorporate farming bill in 1972, Nelson's legislative assistant told environmentalists to support the bill because the "small, independent farmer has close ties to the land and therefore is far superior to the insensitive manager when it comes to environmental protection." The movement toward large-scale irrigation, which had drained water tables in California's Central Valley and was financed by federal reclamation projects, also generated fears among farm leaders about the potential drainage of grain belt acquifers.[24]

Activists included Wendell Berry, probably the best-known critic of "industrial agriculture." When Earl Butz reviewed Berry's book *The Unsettling of America*, it revealed the clash between Butz's view of free-market economic change and Berry's defense of "agricultural fundamentalism" and the "man-earth relationship." The Center for Rural Affairs, founded in 1973 in Walthill, Nebraska, by the activist Marty Strange, consistently advocated reducing the usage of technology and energy-intensive inputs, the usage of which it considered part of the country's "cultural crisis," and promoted "renewable and sustainable" farming as an alternative.[25]

In 1971 NBC produced a program entitled "Leaving Home Blues: An NBC White Paper on Rural Migration." In the program, news correspondent Garrick Utley spoke of "forced migration: the movement of people from rural America who don't want to go. Who would not go if they had a choice. But the choice is gone: devoured by markets and mechanization in agriculture and the failure of industry or government to provide new or adequate jobs." Nebraska, one of the states featured in the program, in the ten years prior to 1971 had seen seventy-three thousand more people leave the state than enter. One Nebraska farmer offered to show the newsmen all the vacated farmsteads and the planted fields where farmers' homes had stood five or ten years earlier. The program forcefully depicted the problem of rural depopulation but infuriated anticorporate farming advocates for not making a more specific link to corporations. The Agribusiness Accountability Project, which helped NBC produce the show, attacked the network's president for his "toothless," "superficial," and "cowardly" production, accusing him of deleting the mention of "every corporate offender that is big enough to cause trouble for NBC." The Agribusiness Accountability Project and others made clear to the public that they believed corporate farming accounted for the outmigration.[26]

Perhaps the most daunting evidence of an impending corporate order in grain belt farming were precedents in the American South, California, and the Third World. The changes in poultry that led to a large degree of contractual integration between corporations and small farmers was discussed the most. As the *New York Times* told the story, "until after World War II, many broilers were raised in the barnyards of family farms. Small flocks of chickens, always underfoot, supplied added income, cash for birthday presents or a winter weekend in the city. Today, there is virtually no market for barnyard chickens. Instead, the family farmer is usually growing broilers under contract for one of the big-agrigiants." The corporation could reduce payments or cut the farmer off completely at any time, but it was difficult for the farmer to escape since he owed the corporation for the production supplies furnished by the corporation. The situation triggered lawsuits on the Delmarva Peninsula and farmer picketing of corporate offices in Northern Alabama.[27] The USDA calculated that the chicken farmers were making about fifty-four cents an hour, an arrangement denounced as "poultry peonage" by Ralph Nader and his raiders.

Roger Blobaum, an Iowa Democrat who ran for Congress in 1970, argued that "the value of corporate secrecy was dramatically illustrated in the 1950s when feed companies persuaded chicken growers in the South to sign contracts that made them as powerless as sharecroppers." Harrison Wellford, a Harvard fellow connected to Nader's Center for the Study of Responsive Law, invoked the fear of such conditions' migrating north: "The role of major national corporations such as Ralston-Purina and Pillsbury in the integrated chicken industry of the south should be instructive for all those who wish, for social or economic reasons, to preserve the independence of the family farmer."[28] The president of the NFO in 1971 issued warnings about the "Kleen Leen" integration contracts offered by Ralston-Purina, who drew "on years of experience turning independent broiler growers in the South into low-income contractors." The president of the NFU had earlier argued that "this is bringing business integration right onto the farm, somewhat reminiscent of the notorious sweat shop system for sending piecework into tenements for cheap hand labor."[29] Jim Hightower asserted that "this corporate invasion of poultry has humbled thousands of these small farmers, reducing them from hearty free-enterprisers to assembly-line cogs." Many feared that the "contract system [that] has turned many chicken producers into little more than low-paid employees of the large broiler companies" would also develop in the grain belt hog, cattle, corn, wheat, and bean sectors. The NFO consistently cautioned against the coming of vertical contracting to the grain belt. As they saw it, packers and processors could use individual contracts with farmers to undermine the collective bargaining for the master contract that the NFO advocated. Ralph Nader came to the NFO convention in 1971 and echoed the complaints about the government's "refusal to invoke the antitrust laws against vertical integration."[30] The contract farming issue took on added prominence in the 1980s and 1990s, and states such as Minnesota and Kansas passed legislation regulating such contracts.

Grain belt fears of corporate takeover were also enhanced by stories originating in California. Tenneco, Standard Oil of California, and Belridge Oil Company, for example, all bought large pieces of land on which to grow fruits and vegetables. The Federal Trade Commission actually charged United Brands and Purex Corporation with seeking to monopolize the production of fresh vegetables, arguing that United Brands was trying to change the lettuce and celery business from one of small, independent growers to one dominated by conglomerates. The concerns among grain belt farmers were also heightened by the steady stream of stories in the postwar period about the plight of farmworkers, especially in California. Many believed that theirs was a future of wage labor to megafarms, similar to the Mexican immigrants picking lettuce, grapes, strawberries, and tomatoes in the Central Valley—like the Joads' journey from independent plains farmers to California farmworkers in *The Grapes of Wrath*. Religious leaders and social reformers in the 1950s, reform politicians and the media in the 1960s, coupled with President Johnson's War on Poverty, highlighted the problems of the farmworkers of California. When the Agribusiness Accountability Project was formed in December 1971, its stated purpose was to study the problems of farmworkers and expose the agribusiness conglomerates that frustrated efforts to help them—it was not long before the mission of the organization was expanded into corporate farming and antitrust areas.[31]

Lurking behind the fears of the California system was a 1946 study—*The Tale of Two Cities*—analyzing two small California towns in the Central Valley, Arvin and Dinuba. One of the communities was dominated by large-scale corporate farming; the other was structured in a pattern similar to rural areas in the grain belt, with the agricultural community comprising many small, dispersed family farms. Although the two towns were in the same climate, produced the same amount of commodities, and were equidistant from other towns, cities, and transportation, other differences were striking. The town surrounded by family farms had more and better schools, churches, recreational facilities, civic organizations, public services, a better standard of living, greater individual ownership, and a 61 percent larger retail trade. Many grain belt leaders believed that such traditions and institutions in their small towns would be destroyed with the coming of corporate farming.[32]

When the United States became heavily involved in foreign aid after World War II, policy makers placed a steady emphasis on land reform in recipient countries. President Truman to the United Nations: "We know that peoples of Asia have problems of social injustice to solve. They want their farmers to own their land and to enjoy the fruits of their toil. That is one of our great national principles also. We believe in the family size farm that is the basis for our agriculture and has strongly defended our form of government." Ngo Dinh Diem received lectures about the need to broaden landownership in South Vietnam, and the countries of Latin America were steered in this direction by the Alliance for Progress, as were all the countries that participated in the United Nations' World Land Reform Conference in 1966. Harold F. Breimyer, an agricultural economist at the University of Missouri, noted that "our development counselors exhort nations so burdened to undertake agrarian reform. Fine; but we ought also be mindful of our own state of affairs." Other farm advocates offered warnings about the potential "Central Americanization" of the grain belt.[33] Later in the same year that Breimyer made his comments, for the first time in American history, a National Land Reform Conference was held in San Francisco and the corporate farming issue discussed. Smaller midwestern land conferences were later held to carry on the work started in San Francisco, and Senator Fred Harris extolled the "need for land reform" in the Senate.[34] The president of the National Farmers Union even advocated a "Land Reform" program in which the federal government would buy good land for resale to small family farmers at reduced prices. In 1975 the NFO called attention to the death of Wolf Ladejinsky, who fled bolshevism in the Soviet Union in 1922, came to the U.S. and received a master's degree in agricultural economics in 1931 from Columbia University, and became an advocate of land reform. By helping to coordinate the redistribution of land in postwar Japan, South Korea, Taiwan, and South Vietnam, he became "Russia's greatest enemy in Asia." The NFO noted that when the communists invaded South Korea, they met little resistance where land was held by a few large landholders but met heavy resistance where land reform had succeeded: "The inhabitants there had a stake in the land and organized to defend it." They juxtaposed Ladejinsky's work with Secretary Butz's "advocacy of big, integrated agricultural operations."[35]

As the Arvin and Dinuba and land reform arguments indicate, one of the core criticisms of corporate farming involved the corrosive affect of concentrated landownership on republicanism and the civic tradition. The images of antirepublican regimes such as the landed caudillos of Latin America, the injustice of concentrated landownership in South Vietnam, the feudal kingdoms of medieval Europe, and the enclosure movement were all invoked against corporate farming. As opponents saw it, the new land barons would be corporate conglomerates like Tenneco and ITT. The goal of the Arvin and Dinuba study was to test the "hypothesis that the institution of small independent farmers is indeed the agent which creates the homogenous community, both socially and economically democratic." The president of the Iowa Farmers' Union was opposed to "corporate agriculture of the Fascist type," and Senator McGovern, on the floor of the Senate, wondered "whether the new society toward which we are heading, a sort of corporate collectivism, is what we really want."[36] Not by accident,

Senator Nelson's subcommittee print after the corporate farming hearings quoted Daniel Webster:

> Our New England ancestors brought thither no great capitals from Europe; and if they had, there was nothing productive in which they could have been invested. They left behind them the whole feudal policy of the other continent.... They came to a new country. There were as yet no lands yielding rent, and no tenants rendering service. The whole soil was unreclaimed from barbarism. They were themselves either from their original condition or from the necessity of their common interest, nearly on a level in respect to property. Their situation demanded a parcelling out and division of the land, and it may fairly be said that this necessary act fixed the future frame and form of their government. The character of their political institutions was determined by the fundamental laws respecting property.... The consequences of all these causes have been a great subdivision of the soil and a great equality of condition; the true basis, most certainly of popular government.[37]

Fewer farms also meant more farmers were forced to migrate to the most unstable, violent (especially in the mid-1960s), socially stratified, and undemocratic of places, the big cities. "The mobs of great cities," Thomas Jefferson said, "add just so much to the support of pure government, as sores do to the strength of the human body." A North Dakota farm couple agreed, arguing that the corporations are "driving contented folks off the land to the already congested, crime-laden city life. This is certainly not the way the Good Lord intended it to be." M. W. Thatcher agreed: "I think and I believe that the most important thing that we have to do to maintain democracy is to preserve on the farm lands that independent husband and wife and those children in that castle on their land, that farm family on their land, supporting these villages and these towns and maintaining that character of life, or you won't have a democracy fifty years from now. It's both, or neither. You don't think there's any democracy in Harlem, do you?" The president of the Farmers Union added other "festering ghettoes—Watts, Detroit, Chicago, and Washington," to the list.[38]

THE LEGISLATIVE RESPONSE

In the 1960s the fears of corporate activity prompted Secretary of Agriculture Orville Freeman to initiate the first comprehensive USDA study of corporate farming. In 1968 USDA researchers found "ample evidence that agribusiness and other nonfarm interests were expanding their investments in agriculture." That same year Senator Nelson, chair of the Small Business Subcommittee on Monopoly, held regional hearings on the subject of corporate farming; one of the stated goals was to update the Arvin and Dinuba study conducted in the 1940s.[39] The first hearing came on May 20 and 21 in Omaha, Nebraska. The national president of the Farmers Union, Tony Dechant, and the presidents of the Nebraska, Iowa, and South Dakota Farmers Unions all testified. Dechant warned of corporate farms taking advantage of tax write-offs, wreaking environmental damage, threatening small rural communities, and interfering with traditional marketing systems. Ben Radcliffe, president of the South Dakota FU, argued that a study by his organization discovered "452 corporations owning agricultural land in South Dakota, totaling 1,633,529 acres, or the equivalent of five medium-sized South Dakota counties . . . one out of every 27 acres of farmland in our state." The second hearing, in Eau Claire, Wisconsin, was delayed by the assassination of Senator Robert Kennedy—who received significant farmer support in his presidential primary wins in Nebraska and South Dakota that year—but produced similar testimony when it was held.[40] Later in the year the NFU released its book *The Corporate Invasion of American Agriculture*, hoping it would publicize the problem and reverse the view that the loss of family farms was inevitable. Senator Nelson saw the problem in iceberg terms, the extent of corporate ownership and control obscured below the surface. Other congressmen predicted that one hundred or so corporations would control all of American farming in the near future. Many agreed and argued that the visible "trends" toward corporate farming were enough to justify alarm and legislative action to prohibit corporate ownership.[41]

Those alarmed by the trends they detected could point to several examples. Based on its experience operating a 180,000-acre ranch in Wyoming, Gates Rubber Company bought several thousand acres in Colorado and also set up egg production operations in Colorado and New Mexico. The Farmers Union argued that Gates was buying so much land that it was driving land values to a level that prohibited local farmers from expanding. Kansas City–based CBK Industries moved into production agriculture, and its president promised a whole new age of farming. The president of the Iowa Farmers Union also reported on the 6,000-acre operation known as Shinrone Farms, which bought hundreds of thousands of dollars in machinery from Massey-Ferguson and painted the machines white with green shamrocks owing to the "sentimentality about Ireland" of the owner, a Detroit trucking executive. The Center for Rural Affairs reported on forty-three "factory-type" hog operations in Nebraska in 1974. Environmental Applications Inc. started an eighty-acre operation in southwestern Minnesota to produce thirteen thousand hogs a year; Swift and Ralston-Purina proposed a 2.5 million-a-year hog production operation near Kahoka, Missouri, which the NFO feared would eventually "monopolize a whole type of production" through "elimination of independent producers of hogs and consequent elimination of effective and efficient competition"; the Arizona-Colorado Land & Cattle Company expanded its holdings to 1.2 million acres and started cattle feedlots and meatpacking; the Ceres Land Company of Sterling, Colorado, acquired several thousand acres of land in eastern Colorado to irrigate pastures and feed cattle. More familiar companies cited in the debate include Tenneco, American Cyanamid, Bunge Inc., Del Monte, Goodyear, Gulf & Western, Heinz, Libby-McNeil & Libby, Minute Maid, Pillsbury, Standard Oil, DuPont, Dow, Chase Manhattan, Getty Oil, and Textron. Tenneco, the thirty-fourth largest U.S. corporation, which promised its stockholders "integration from seedling to supermarket," became a symbol of the much-feared "corporate farmer." Once Tennessee Gas Transmission Company, Tenneco expanded into oil production, shipbuilding, chemicals, manufacturing (including the acquisition of the farm machinery manufacturer J. I. Case), tens of thousands of acres of land for growing fruits and vegetables, and was seen as representative of the "conglomerate invasion of agriculture."[42]

The NFO became particularly adamant about restrictions on corporate farming. The collective bargaining program for farmers they advocated could work only if packers and processors were prevented from integrating backward and producing their own livestock and grain. When a mandatory bargaining framework for farmers similar to the National Labor Relations Board for workers was debated in the later 1960s and early 1970s, some farm advocates feared such an arrangement would drive processors into production themselves. When criticizing the legislation, NFO president Oren Lee Staley argued that the most urgent need was for "legislation to prevent handlers, processors, integrators and suppliers from moving into agricultural production to escape bargaining."[43]

In the early years of the debate the studies and statistics of corporate farming were always in dispute. At the opening of Senator Nelson's hearings in Omaha, two agricultural economists from the University of Nebraska argued that information on the level of corporate farming was simply not available, despite the examples cited by the many witnesses calling for legislation. The NFO, on the other hand, constantly pointed out that the level of contractual integration was obscured, that the USDA was using 1963 statistics as late as 1971, and that professors at land grant colleges were afraid to conduct research properly because they feared losing grants offered by companies such as Ralston-Purina and Safeway.[44] In 1971 Senator Nelson held congressional hearings on agribusiness "secrecy" in an attempt to get better statistics on the number of corporate farms in operation. The chairman of the Agribusiness Accountability Project, Phillip Sorenson, called for a law requiring the Securities and Exchange Commission to ask companies about their farming activities and for an annual report to Congress on corporate involvement in agriculture.[45]

In 1969 the agricultural census included corporate farming for the first time. In the first compilation of data from the census, released in 1972, the Census Bureau reported only 21,513 farms with sales over $2,500 that were "corporate," or about 1.2 percent of the total number of farms. Of these farms, 19,716 had less than ten stockholders, indicating many incorporated family farms. But several of the conclusions of the researchers were challenged by Professor Richard Rodefeld, a sociologist at Michigan State, in a speech to the First National Land Reform Conference. Rodefeld argued that early statistics counted the sharecroppers on large southern plantations as part of a "multiple-unit operation," but later statistics counted the sharecroppers separately, giving the impression that fewer large farms had appeared and that fewer family farms had disappeared. He also argued that the aggregate statistics obscured the large-scale corporate involvement in states such as California and Arizona and that the amount of sales accounted for by large farms was grossly disproportionate to their actual numbers. Senator McGovern argued that the "new information" proved that he and other anticorporate farming advocates were not "crying wolf" and called on the USDA to stop ignoring the problem; USDA economists responded by calling the "corporate farmer" a "straw man."[46]

The statistical clash was showcased in the early 1970s when Congress held hearings on corporate farming. The American Farm Bureau Federation told Emanuel Celler (D, NY), chairman of the House Judiciary Committee, that there is "little solid evidence that [the entry of conglomerate corporations into farming] is a serious problem." On the basis of the best USDA statistics available, they argued that less than 1 percent of farms were incorporated and that many of these were family owned. J. Phil Campbell, the undersecretary of agriculture representing the Nixon Administration, also opposed legislation limiting the corporate ownership of farmland on this basis, while also fearing the damage it would do to existing corporate farm operations. The National Grange also worried about the precedent, arguing that "it would be the first time in our knowledge that Congress had passed legislation that would limit by law persons who could engage in a particular industry." Professor Rodefeld testified that aggregate statistics were misleading, that corporate farms were actually dominant in certain parts of the country, and that the "inevitability" of corporate farming justified early legislative action. He challenged the USDA on what he considered their incompatible views—corporate farming was not a problem, but an anticorporate farming law would hurt many corporations.[47]

The intense fears associated with corporate farming would slowly subside as more and more studies indicated that the "invasion" argument was overdrawn. The Food and Agricultural Act of 1977 included a provision directing the secretary of agriculture to conduct a large-scale investigation of the structure of agricultural production. The Carter administration's new secretary of agriculture, Bob Bergland, a farmer from northern Minnesota, announced a "national dialogue" on the structure issue at the National Farmers Union annual meeting in 1979. He aimed to overcome the earlier statistical problems by "amass[ing] the most comprehensive and reliable base of data ever compiled." By 1982 the USDA numbers indicated that corporate farms still represented only 2.6 percent of all farms in the country but accounted for 23 percent of product sales.[48]

Whatever the statistics, the concern about outside corporate control was long-standing in the grain belt, and unlike other regions in the 1970s, a consensus existed that the trend toward corporate farming should be opposed. The Populists had passed laws limiting corporate farming in Minnesota and Nebraska, and similar laws were passed in North Dakota and Kansas in the 1930s. Building on this tradition in the 1970s, South Dakota, Minnesota, Wisconsin, and Missouri limited corporate farming, and Iowa and Nebraska adopted formal reporting requirements. Iowa also limited ownership of land by trusts other than authorized farm trusts, family farm trusts, or testamentary trusts. The populist antagonism for concentrated economic power and the agrarian proclivity for dispersed and decentralized ownership were still powerful in the grain belt and evident in the laws passed in the 1970s. The Iowa law spelled out the need for a dispersed agriculture and the need to preserve small communities; the Nebraska law was designed to prevent the potential monopolization of agriculture; and the

South Dakota law "recogniz[ed] the importance of the family farm to the economic and moral stability of the State."[49]

Starting in 1971 the U.S. Congress also held several hearings on bills—usually entitled the Family Farm Antitrust Act—which would have formally outlawed ownership of farmland by anyone possessing more than a few million dollars in nonfarm assets.[50] The federal efforts were consistently stymied, however, by the thin evidence of an "invasion," the reluctance of the USDA and the Department of Justice, the criticism of the measure for midwestern "parochialism," and the fact that Congress, unlike farm state legislatures, was not made up of farmers and small-town business leaders dependent on the farm economy. Political concerns about the size of the bonanza farms on the nineteenth-century plains eased when the farms "disintegrated in the hard nineties," and so too did worries about corporate farming ease when large-scale farm operations such as Black Watch Farms and the farm operated by Gates Rubber Company failed, further deflating the legislative efforts.[51] Numerous studies indicated that larger farms were not more efficient than smaller, family farms—maximum cost-saving production efficiency could be reached farming under one thousand acres of corn and wheat.[52] It was also discovered that the reason some corporations became interested in farming in the 1960s and 1970s stemmed from a tax loophole that was subsequently closed. The federal legislative effort also sputtered because many farmers benefited from acquiring new land and, in the Midwest, most consolidation stemmed from farms being bought by other farmers, not by outside corporations. The backdrop for Jane Smiley's Pulitzer Prize–winning novel *A Thousand Acres*, for example, is an Iowa farmer's arguments with his neighbor "about who should get the Ericson land when they finally lost their mortgage."[53]

Even Ben Hogan's nemesis, the Nowell-Safebuy, went bankrupt. The company should have noticed the "Bates Rubber Company's fiasco" with wheat farming in Colorado. Bates had to hire 50 percent more workers than the total number of independent farmers who previously ran the farms for the same amount of production. Safebuy, soon after it adopted its corporate ownership strategy, began losing money, laid off its workers at the turkey processing plant, and shut down. Although the extent of corporate ownership of American farms was exaggerated, worries about the social changes in rural life underlying the corporate farming debate reflect deeper concerns about a republic with fewer farmers and fewer small towns.[54] While the threats of such companies taking over American farming may have been overdrawn at times, the contentiousness of the debate indicates the degree to which many farmers and legislators valued a system of independent agricultural producers. The contentiousness spilled over into suspicions of monopolistic practices by meatpackers, grain processors, and grain traders, who, many believed, were undermining the economic sustainability of family farms.

3

THE POLITICAL ECONOMY OF MEATPACKING AND GRAIN PROCESSING

> It is very unfair competition for the farmers because again they will pay us what they want to. Again they (large packers) control the price of what we get paid and there is to much of a margin between the farmers price and the meat market price to the consumer. MR. AND MRS. HERBERT HERREN, FARMERS, MONTICELLO, IOWA (PARENTHESES IN ORIGINAL)

Throughout the 1940s and 1950s the food industries received reduced attention from antitrusters. From the time of the Temporary National Economic Committee's (TNEC) monograph on the food-processing industries until a massive Federal Trade Commission study of the food system in the 1960s, few economists focused on questions concerning the industrial organization of the food processing sector. James McNicholls, who published his book *A Theoretical Analysis of Imperfect Competition with Special Application to Agricultural Industries* in 1941, left the academy in the early 1950s when "unable to convince many agricultural economists to apply the industrial-organization framework to study the food system."[1] Constant questions about corporate farming and charges of monopoly from farm groups such as the NFO and the NFU, however, prompted greater attention from economists and government officials in the early 1960s. Congress funded the massive studies of the National Food Marketing Commission—which the NFU asked to be "as broad as the TNEC"— and by the mid-1970s the FTC completed another ten studies of the food industry. The "McGovern Papers," FTC inquiries leaked to the McGovern campaign in 1972, were used as evidence of widespread monopoly "overcharges" by food processors. In addition, inflation prompted the Council on Wage and Price Stability to release studies and Congress to hold thirty-one days of hearings on the food industry between 1973 and 1976, calling over 250 witnesses.²

The large-scale studies and investigations, similar to those pertaining to corporate farming, failed to provide a smoking gun justifying more stringent antitrust action or government-imposed divestiture of food firms. Food firms did become larger and economically stronger, however, underscoring the need for farmers to organize and market their products effectively. And when the "intellectual foundations of antitrust . . . crumbled and collapsed" in the 1970s, reducing the possibility of state-imposed deconcentration in the food industries or of maintaining smaller food firms through a strict merger policy, the need for farmers to organize more effectively increased further. Nevertheless, throughout the postwar period the sectors farmers sold their livestock and grain to showed signs of competitiveness.³

The low level of concern about anticompetitive practices in food manufacturing stemmed in part from its historic structure. Some have argued that the series of small- and medium-sized firms that made up the food sector before World War I made it the "most competitive major sector of the economy." In the 1920s, however, mergers prompted by economies of scale, vertical integration, and the emerging medium of national advertising campaigns created some large food corporations "almost overnight." During the Great Depression and World War II this growth stalled, but in the postwar period the monopoly concern became more pronounced as the number of food manufacturers dropped by over 50 percent from 1947 to 1972. In the mid-1960s "an avalanche of mergers broke loose in the U.S. economy"—"merger mania"—and from 1971 to 1975 food- and tobacco-manufacturing firms made 25 percent of all large manufacturing acquisitions. A. C. Hoffman, an early pioneer in the field of competition in the food industries, claimed that "never before in the history of capitalism [had] such great aggregations of economic power been created."⁴

Concentration in food processing was a matter of not simply product concentration but aggregate concentration. Food processing transformed itself from "being the most specialized of manufacturing firms in 1950" to "among the most diversified." In the late 1970s the "average" food processor earned 60 percent of its sales outside its principal industry, and by 1981 the top fifty food manufacturing companies averaged 8.7 different areas of manufacturing. Firms that mainly focused in other areas, such as chemicals and food wholesaling and retailing, also increased their food manufacturing output. The conglomeration of production became a core

TABLE 2. THE DIVERSIFICATION OF ARMOUR AND COMPANY

Nonmeat foods	Chemicals and industrial products	Agrichemical products	Heavy equipment	Pharma-ceuticals	Household grocery products
Butter and cheese	Fatty acids	Fertilizers	Turbines	Hormones	Soap
Vegetable oils	Nitrogen	Ammonia	Presses ship props	Drugs	Floor wax
Margarine	Abrasives	Insecticides	Pumps	Veterinary supplies	Glycerine
Shortening	Shoe leather	Fungicides	Desalination equipment		Detergents
Pet foods	Derivatives	Phosphates			
		Weed killers			

Source: Richard Julius Arnould, "The Effects of Market and Firm Structure on the Performance of Food Processing Firms" (Ph.D. diss., Iowa State University, 1968), 86.

characteristic of the postwar economy, especially after the Celler-Kefauver Act of 1950 limited horizontal mergers and fears of a return to depression drove efforts to diversify, a strategic move designed to increase managerial flexibility during economic downturns. By the 1960s only one-eighth of mergers were horizontal—most of the rest were conglomerate.[5] Conglomeration was not simply a national phenomena—the firms created held subsidiaries and divisions all around the globe.[6] Armour and Company offers an example. From a processor of meat and meat by-products in the 1920s the firm expanded by 1966 into all of the areas indicated in table 2.

The problem of conglomeration proved to be one of the unforeseen weaknesses in postwar antitrust efforts. Brown Shoe, the first postwar merger case to make it to the Supreme Court, solidified the enforcement of the Celler-Kefauver merger law in horizontal contexts—from 1962 to 1970 the Supreme Court decided twenty-eight out of twenty-nine merger cases in favor of the government. Along with the Von's Grocery decision, in which the Supreme Court prohibited firms with very small market shares from merging, the courts "virtually [stopped] all but very small mergers by the leading ten food chains."[7] The first horizontal mergers challenged after Celler-Kefauver involved food companies—the FTC's first case involved Pillsbury Mills, and the Department of Justice's first case involved Schenley Industries Inc. (liquor). Conglomerate mergers instead became the greatest source of concentration, and as a result, some were declared illegal. Conglomerates generating concern in the food industry included Proctor and Gamble, which moved into various food manufacturing and cleaning product sectors and by 1982 sold $12 billion in products. In 1967 the Supreme Court upheld an FTC decision to divest P&G of Clorox Chemical Company. International Telephone and Telegraph also sought to diversify beginning in 1960—by 1981 it expanded by over 2,000 percent, including the acquisitions of Avis, Sheraton, Hartford Fire Insurance, and Continental Baking Company, the largest baker in the United States. In 1971 the government abandoned three antitrust cases against ITT, eliminating the possibility of a broad Supreme Court ruling on conglomerate mergers, especially after the emergence of the court's "new antitrust majority" in 1973–74. Even the rules governing horizontal mergers were subsequently weakened, contributing to the "record volume of food manufacturing acquisitions" in the early 1980s.[8] These changes in the structure of food processing, coupled with the postwar depopulation of rural America, periodic farmer aggravation with commodity prices, greater corporate interest in agricultural production, and the continuing farmer interest in antitrust in the 1960s and 1970s, perpetuated the serious concerns about corporate power and the monopoly problem.

Despite these mounting concerns, it is not possible to indict the grain belt food-processing sector for a complete absence of competition. Rival firms at times talked of "waging war" with one another, the demise of uncompetitive firms was widespread, and many of the changes in plant size and concentration level could be justified by economies of scale. The conglomeration of agribusiness, while generating some additional efficiencies, did not always generate greater-than-normal profits and is increasingly seen as part of a failed experiment in conglomerate growth in the postwar era.[9] Still, the relinquishment of antitrust activities in the 1970s and early 1980s together with the conglomeration of food processing remains a concern, given the relative organizational position of farmers. While continuing a stringent antitrust policy probably could not have been justified on purely anticompetitive grounds, it might have been made on other grounds, such as promoting the relative bargaining power of the farm sector and a decentralized vision of the American economy.

MEATPACKING

In the late nineteenth century many farmers suffered through the consequences of deflation and falling demand for beef. Many farmers thought their problems stemmed from the growing market power of Chicago meatpackers, whose new variety of low-cost, refrigerated, dressed beef was spreading around the country. Local slaughterhouses and packing firms, numbering close to a thousand in 1880, could no longer compete with the new product. Some studies estimate that beef carcasses could be shipped from Chicago at one-third the cost of shipping live cattle to packers east of Chicago. As a result, these firms joined the protesting farmers, sometimes claiming that the Chicago beef was unhealthy and demanding government inspections of meatpacking plants.[10]

The lower slaughter costs and the growing consumer access to refrigeration, which increased demand for beef, prompted the "cattle boom" of the early 1880s, which ultimately led to greater supplies and falling prices in the late 1880s: from 1884 to 1891 cattle prices dropped from $25.26 to $16.49 per hundredweight. These price changes, combined with farmers' protests about railroads and other trusts and the complaints of smaller packers, fed fears about "big four" meatpacker collusion and contributed to efforts to pass national antitrust legislation. During the debate over the Sherman Act, Congressman Richard Bland of Missouri argued that "there is no trust in this country that today is robbing the farmers of the great West and Northwest of more millions of their hard-earned money than this so-called Big Four beef trust of Chicago."[11]

After a period of rising livestock prices in the 1890s and early twentieth century, monopoly fears reignited. An investigation from 1917 to 1919 by the newly founded Federal Trade Commission concluded that the meatpackers' market power and efforts to diversify indicated that they were attempting to "monopolize" the nation's food supply. The resulting 1920 consent decree restricted the big meatpackers from owning any interest in stockyards, terminal railroads, or market newspapers or journals, from using their networks to deal in nonmeat products or owning a controlling interest in nonmeat products, and from retailing meat and owning warehouses. In 1932 the Supreme Court ordered the major packers to comply with the decree. Despite all the other restrictions, the government "left the packers more or less unencumbered with respect to meat packing."[12]

In the postwar period the American meatpacking industry changed dramatically, rendering the consent decree meaningless. Most of the older firms that dominated the industry in earlier decades and prompted so much public scrutiny went under or were acquired by ascendant firms or conglomerates. Economic pressure also stemmed from concentrating power in the feeding and retailing sectors, diminishing per capita demand, competition from such other meats as chicken, pork, and fish, a growing level of imports, the emergence of a series of new firms with plants closer to beef supplies that could take advantage of lower labor costs, a productivity "revolution," and a degree of farmer organization. The many structural changes in the industry also indicate its dynamism.

One of the checks on packer market power was the growing power and concentration of the feeding industry. Previously, much of the final feeding of slaughter cattle took place on small farms mostly located in the grain belt. In 1962 two-thirds of cattle marketed for slaughter came from feedlots with a capacity of less than 1,000 head, while feedlots with a capacity between 1,000 and 1,600 head largely composed the other one-third; by 1973 over two-thirds of the cattle were fattened on large commercial feedlots—20 percent of cattle were fattened on feedlots with a capacity greater than 32,000 head. The thirteen counties of the North and South Platte River basins, for example, saw the number of farmer-feeders (500 head or less) shrink by one-sixth from 1953 to 1959, while the number of commercial feeders more than doubled. By 1964 Nebraska led all other states, with 830 feedlots with at least 1,000 head of cattle. Since many of the new operations were located in the central and southern plains, as opposed to the more traditional corn belt, Iowa Beef Processors co-chairman Currier Holman warned Senator Dick Clark (D, IA) that Iowa risked "losing" its feeding industry. By 1982, 381 feedlots with a capacity of at least 8,000 head marketed over half of the fed cattle in the twenty-three leading states, and by 1995 over 90 percent of beef cattle were marketed from feedlots with a capacity of at least 1,000 head, prompting packer competition for access to the larger suppliers. Owing to packer competition for large blocks of cattle, an internal USDA memorandum in the 1960s estimated that larger feedlots could receive as much as twelve dollars more per hundredweight from packers.[13]

The increasing importance of the larger feedlots meant that more and more cattle were sold directly from the feedlot to the packer, bypassing the terminal market stage. From 1950 to 1964 the percentage of cattle that packers bought on terminal markets dropped from 75 to 37 percent, and the percentage of hogs from 40 percent to 24 percent. In 1953, 78 percent of the cattle sold in Colorado were sold at the terminal markets, for example, but by the early 1960s, 66 percent of the cattle slaughtered in the Denver area traveled directly from feedlot to packer. In 1984 all packers purchased roughly 79 percent of their beef supplies directly from feedlots, 7 percent from terminal markets, and 15 percent from auction markets. Direct packer involvement in feeding was always quite small and shrinking: by 1980–82 the percent of packer-fed cattle dropped to 4 percent from a level of 7 percent in the 1960s.[14]

The food retailing sector also emerged as a powerful constraint on the market power of the packers. At the time of the large-scale FTC investigation from 1917 to 1919 the packers operated their own branch houses and sold 93 percent of all fresh and cured (wholesale) meat. As a result, grocers could not "exert any substantial influence on the major packers." After the consent decree prohibited packer involvement in retailing and the coming of chain retailers, this relationship changed dramatically. Grocery stores with meat increased their share of the market from 31 percent in 1929 to 74 percent in 1954, and from 1920 to 1958 the percentage of grocery sales accounted for by chains and cooperatives jumped from 11 to 85 percent. The power of supermarkets and food chains, some of whom started to pack their own beef, even sparked concerns about a "new monopoly." Some farmers even criticized chains like Safeway, Kroger, and A&P more than they did the packers. Mary Yeager notes that "in this changed world of retailing, where buyers had the power to choose their suppliers and enforce competitive bidding, the large packers had no special advantages over the smaller, more specialized packing concerns."[15] A&P's success at securing lower prices from food manufacturers in the late 1940s and Rath's and Iowa Beef Processors' (IBP) jockeying for the business of Safeway in the 1970s indicate the growing influence of food retailers. The FTC studies in the 1960s took note of this growing power, and in the 1970s, during House Small Business Committee hearings, one economist noted that countervalence advocates such as John Kenneth Galbraith would applaud the changes in both food retailing and cattle feeding and the constraints they placed on packers.[16] In the 1970s and early 1980s, farmers attempted a large-scale antitrust suit against grocery chains for depressing meat prices, but the suit failed owing to complications relating to legal standing.

A further check on the market power of the food-processing sector stemmed from the power of restaurants. In 1967, 31 percent of total food sales were classified as "away-from-home," including "public eating places," schools, hospitals, colleges, the military, and others. The power of food retailers such as restaurant chains, coupled with independent and chain groceries, would place steady pressure on food processors to keep prices competitive. Part of the reason for the minimal backward integration into food processing by chain grocery stores is the modest profit margins in parts of food processing and the already commanding market power of the grocery chain stores. Competition increased, however, by the degree to which grocery chains decided to enter the food manufacturing sector. In the mid-1960s, 8 to 9 percent of food sales were from products manufactured by grocery chains—Safeway owned three meat-processing plants, for example. This number, it would seem, is not higher because the food-processing sector exhibits general indications of competitiveness and therefore did not promise extracompetitive profits to grocers who decided to integrate backward.[17]

The tenuous aggregate demand for meat also constrained potential packer power. In 1950, for example, per capita beef consumption was 71 pounds; when inflation cut purchasing power the next year, consumption declined to 63 pounds. In 1955 consumption reached the "gastronomical feat" of 91 pounds, but an intervening urban recession reduced the level of consumption to 81 pounds by 1959. Consumption of beef peaked in 1976 at 127 pounds a year—twice the 1950 level; during the economic stress of 1979–81 it dropped to 104 pounds. A manager of economic research at Wilson and Company noted how the changes in demand were closely linked to income and inflation.[18] The manager of the Grain Terminal Association's durum wheat mill admitted that when beef prices increased the consumption of macaroni increased, adding that his "competition [was] not among [wheat millers], but with other foods and the rapidly changing habits of the American people." Beef increasingly competed against chicken and fish as reports about carcinogens in beef and the links between heart disease and strokes and red meat consumption increased. Senators from livestock-dependent states such as Carl Curtis (R, NE) attacked Giant Food Inc., who ran advertisements ridiculing the price of red meat and promoting such items as "boneless shad." Curtis tried to interest the FTC in Giant Food, since the company charged more for "boneless shad" than for shad fillets, even though there was no such thing as "boneless shad." His office also pointed out that Giant charged "excessive prices for rock fish," since they sold at $.40 pound wholesale but $1.89 "whole or uncleaned," a 450 percent markup, all to discredit a product that Curtis and others feared would be substituted for beef. By 1987 it seemed that Curtis had lost his battle, as per capita consumption of chicken actually surpassed beef for the first time and some predicted that beef consumption would decline to 50 pounds per year by the year 2000.[19]

The "shock" in the demand for beef was so severe that from 1979 to 1986 real beef prices had to decline over 30 percent to maintain a constant level of beef sales. As a result, beef herds were liquidated—from about 132 million head in 1975 herds shrunk to about 100 million fifteen years later, creating a 20–30 percent excess capacity in packing plants. Pork also suffered from demand problems. In the early 1970s a vice-president at Oscar Mayer noted that in the late 1940s pork commanded 40 percent of the red meats and poultry market but slipped to 28 percent by 1970. The inflation of the early 1970s also triggered protests about food prices—which rose 15

TABLE 3. DECLINING CONSUMPTION AND PRICES OF MEAT, 1970–1989

	Per Capita Consumption (lbs)	Deflated Retail Price ($ per lb)
1970	84	2.62
1980	76	2.88
1985	79	2.16
1989	69	2.14

Source: Wayne D. Purcell, "Economics of Consolidation in the Beef Sector: Research Challenges," *American Journal of Agricultural Economics* 72 (Dec. 1990): 1212.

percent in 1973—meat boycotts, and the organization of a consumer movement quick to criticize what they viewed as high food prices and to urge substitute products for meat.[20]

Domestic meatpackers also faced more and more competition from imports in the postwar period. In 1955 beef and veal imports represented 2 percent of domestic supply but expanded to 9 percent by 1963.[21] Farmers shared packer concerns about imports to the extent that they outweighed concerns about packer fair play. In 1952, according to John T. Schlebecker, even a "piddling 40 million pounds of commercial grade beef" from New Zealand "offended the delicate sensibilities of the cowboys." But from 1958 to 1963 alone, imports of Australian beef increased from 18 million pounds to 517 million pounds. Between 1960 and 1963 total meat imports increased by about 900 million pounds, depressing hamburger prices at least 10 percent.[22] Farmer concerns about the price drop explains the passage of the Meat Import Act of 1964, which reduced imports 8 percent in the law's first year and made import restrictions the primary method of livestock price support by the federal government. For the next decade the government cobbled together an import program mixing quotas, voluntary export restraints of foreign traders, and open trade in an attempt to manage market conditions. Cattle producers were often furious with the results, which they felt were too often manipulated by the State Department and not the USDA. Various White Houses also paid close attention to the connection between trade and meat prices, one taking particular notice of how high meat prices contributed to the defeat of Prime Minister Harold Wilson in England in 1970. Richard Nixon: "We can't let that happen to us."[23]

The extent of imports remained controversial—as one farmer put it, the free-trading "boys [weren't] too popular around the livestock barns" in those years. Competitive pressure continued, and by the 1980s imports still totaled about 9 percent of commercial production (Australia accounted for about one-half, New Zealand about one-quarter, and Argentina, Canada, and Brazil rounded out the total). Exports totaled only a little more than 1 percent of commercial production (most of which went to Japan), even though total exports had grown from $27 million in 1965 to $392 million by 1983. In 1988 the chance of greater export growth was further constrained when the EEC banned American beef raised with growth hormones. Greater access to world markets, favorable exchange rates, and lower prices boosted exports to 7 percent of total production in 1995, however, nearly equalizing the amount of imports and exports, and within the next seven years one of the leading cattle slaughterers hopes to boost foreign sales to 25 percent of its total sales.[24]

At times, packers were also forced to compete for cattle and hogs supplied directly by farmers. The National Farmers Organization, which organized holding actions in the 1950s and 1960s in order to boost farm prices, succeeded in disrupting packers' traditional supplies. The Hormel plant in Austin, Minnesota, for example, lost local supplies to packers in other cities owing to the NFO. In the spring of 1968 Rath Packing reached an agreement with the NFO that it would pay a "local country price" for hogs originating from the NFO's country collection points plus fifteen cents—five cents for arranging the delivery and ten cents for the collection point operators—for twelve hundred hogs daily. The packers who signed contracts with the NFO, hoping to reduce procurement costs, may have suffered a competitive disadvantage in the long run. The Morrell plant in Ottumwa and the Rath plant in Waterloo signed NFO contracts and ultimately sold out or closed down. IBP, which did not, successfully acquired many of the failed packers' plants. A Hygrade official noted the competitive disadvantage of NFO contracts shortly before they sold out to IBP: "What we can't afford to do is pay more than our competitors." The organization of farmer marketing in various places around the grain belt created unequal supply costs and therefore compounded the competitive pressures on many packers.[25]

Farmers and farm organizations also went beyond price bargaining and started cooperatively packing hogs and cattle, as in the case of Farmland Industries. Financing was available from the Banks for Cooperatives, technical advice was supplied by the Farmer Cooperative Service, and costs weren't prohibitive, according to Farmers Union estimates.

Packer profits could also give some indication of market power. But such a measure is complicated by the practice of "cross-subsidization," or the

TABLE 4A. ESTIMATES OF COST OF BEEF-SLAUGHTERING PLANTS OF VARYING CAPACITY (IN THOUSANDS OF DOLLARS)

Annual Animal Capacity	Slaughter Only	Primal Cut	Inedible Processing Breakdown	Operating Capital	Total
5,000	150	5	52	207	414
10,000	150	5	40	102	297
37,500	375	9	75	375	834

TABLE 4B. ESTIMATES OF COST OF HOG SLAUGHTERING PLANTS OF VARYING CAPACITY (IN THOUSANDS OF DOLLARS)

Annual Animal Capacity	Slaughter Only	Cutting Only	Inedible Processing Breakdown	Operating Capital	Total
75,000	213	25	75	313	626
150,000	350	25	40	155	570
300,000	500	35	60	325	920

Source: Division of Community Development Services, "New Opportunities for Farmer Cooperatives," Dec. 15, 1964, FF 21, DB 1, ser. 3, NFU Papers, UCB.

shifting of resources from division to division in diversified conglomerates, obscuring actual costs and margins for specific products—or, simply put, ITT counting Twinkies as telephone equipment. In the 1960s Ray Goldberg, a prominent agribusiness analyst, admitted that "the profitability of various segments of firms in selected industries must be arbitrarily extracted from integrated operations." Even Joan Robinson admitted that the complexity of the postwar firm threatened to make much of her work in the area of industrial organization obsolete: "My old-fashioned comparison between monopoly and competition may still have some application to old-fashioned restrictive rings [cartels] but it cannot comprehend the great octopuses of modern industry. . . . More and more, the great firms have a foot not only in many markets but in many industries, in several continents."[26]

In one comparison of industry-derived statistics and USDA-derived statistics, an effort was made to understand the problem of determining profit margin in an integrated firm, in this case flour milling. The industry-derived statistics indicated that within the firm the flour-milling operation lost 0.33 percent on equity after taxes. The USDA statistics, which did not segregate flour milling from the other operations of the diversified firm, indicated an 8.4 percent return on equity after taxes. A comparison between agribusiness and USDA data for margarine also indicates pitfalls. The agribusiness data involved only chain stores, but the USDA data involved a nationwide average of all retail outlets. The chain store margins were 13 percent smaller than the aggregation of all retail sales. The National Food Commission found that the USDA consistently overstated the price of beef because it didn't take into account sales, specials, discounts, and other merchandising efforts. These accounting problems underscore the "complex and diverse patterns that emerge when one merely examines the different merchandising policies of the selected retailers." After the Food Commission submitted its report, the head of the Department of Agricultural Economics at Purdue argued that "[the] report begs for greater sophistication of bench-mark data" and specifically criticized sloppy USDA margin studies. The failure to prove a high degree of competition because of the complexity of the accounting problems makes another point, however. If researchers can't figure out the complexities of the costs, margins, and variable marketing strategies after the fact, would-be colluders probably couldn't either—reducing the chances that they could successfully police a collusive arrangement.[27] An absence of concrete data, however, also means that collusion was possible, justifying the concerns of farmers and antitrust enforcers.

Perhaps the most compelling evidence of competition in the meatpacking sector stems from the structural "re-engineering" of the industry after World War II. In 1950 the concentration ratio of traditional "big four" firms stood at 52 percent, but by 1972 this number had dwindled to 25 percent. The "big four" oligopoly shattered after the war with the growth of the "little five" (see tables 5a and 5b). Other, more cost-efficient rivals such as Iowa Beef Processors and MBPXL also emerged. By 1982 the top-four concentration ratio crept back to 35 percent but involved new firms. As the leadership of IBP argued, the "entire industry [had been] revolutionized!"[28]

The shake-out triggered many plant closings and many industry mergers—between 1967 and 1982 the number of meatpackers with twenty or more employees dropped from 955 to 668. In 1967 the conglomerate Ling-Temco-Vaught acquired Wilson and Company and decided to close many of its beef plants and focus on pork. The shutdown included plants in Albert Lea, Cedar Rapids, Oklahoma City, and Louisville. Armour, the second largest meatpacker after World War II, was acquired by Greyhound Corporation in 1970 and made a failed attempt to expand into fast food. Armour ended up closing plants in Sioux City, Green Bay, Omaha, Chicago, Brownsville, and Fort Worth—from 1956 to 1963 alone

TABEL 5A. TOTAL SALES OF BIG FOUR (IN MILLIONS OF DOLLARS)

	Swift	Armour	Wilson	Cudahy	Total
1952	2,593	2,184	826	563	6,166
1962	2,495	1,859	711	313	5,378

TABLE 5B. TOTAL SALES OF LITTLE FIVE (IN MILLIONS OF DOLLARS)

	Oscar Mayer	Hormel	Morrell	Rath	Hygrade	Total
1952	217	306	292	253	137	1,205
1962	270	385	571	273	456	1,955

Source: Lawrence A. Danton, "The Decline of an Oligopoly: Changes in the Meat Packing Industrial Structure," *Rocky Mountain Social Science Journal* 5, no. 1 (Apr. 1968): 43–44.

it closed sixteen plants.[29] Hormel, another one of the nation's largest meatpackers, leased its slaughtering plants and attempted to diversify into a "food company." Hormel succeeded in reducing its dependence on meat, and its ratio of branded, value-added products rose from 30 percent of total sales in 1979 to 75 percent by 1989. In 1980 Esmark announced the selling of most of its Swift subsidiary, including almost all the beef operations. The plants that closed represented 70 percent of Swift's steer and heifer slaughter. This move included selling a plant in Grand Island and closing plants in Des Moines, San Antonio, Rochelle (Illinois), and Guyman (Oklahoma). In 1980 Monfort closed its Greeley plant, where only two years before it conducted its entire beef slaughter. Also in 1980 Farmland Industries closed its 285,000-annual-slaughter plant in Garden City, Kansas. John Morrell was acquired in 1967 by AMK and six years later closed its Ottumwa plant—original home of the company's headquarters—and teetered on the brink of closing its Sioux Falls and Sioux City plants for years. Such changes prompted Dick Knowlton, head of Hormel, to comment that "people talk about the steel industry being devastated. What happened to the meat industry makes that look like a Sunday school picnic." A House Small Business Committee report concluded in 1980 that "there are few industries in the nation today which are in such turmoil."[30]

The most successful and aggressive of the new firms was Iowa Beef Processors, a company that started with the help of a 1960 Small Business Administration loan in Denison, Iowa. IBP often bid up the price of live cattle in an area to divert supplies from the older packers. As one former IBP official acknowledged, "the goal was to be the lowest cost producer [and] the lowest cost slaughterer, to enable us to pay a quarter a cent a pound more for desirable cattle and take them away from the competition." IBP also bought large quantities of carcasses from the older packers, waited for their distribution systems to erode, then stopped buying from them; unable to outbid IBP for live cattle, and without a solid distribution system, the older firms often folded. In 1975, 45 percent of Dubuque Packing's carcasses went to IBP, as did 25 percent of Hyplains' (Kansas), 30 percent of Midwestern's, 25 percent of Platte Valley's (Nebraska), and 50 percent of Amarillo Packing's. Many of these carcasses were further processed into "boxed beef," or packer-packaged individual cuts of meat ready for sale in grocery stores. IBP's share of boxed beef sales approached 50 percent of total sales in the mid-1970s.[31]

IBP also tried to increase market share by offering discounts to large retailers, a risky strategy given the Justice Department's quick action against John Morrell and Company for offering gifts to retailers in 1965. At about the time Rath Packing Company was seeking Safeway's business, IBP made Safeway a rebate offer: $.50 per hundredweight if they bought 500 cattle per week; $.75 for 750; $1 for 1,000. IBP officials knew they risked Robinson-Patman Act violations for price discrimination for not offering discounts to distributors also. Aside from potential objections from distributors, they knew that any discounts required a cost-saving justification and therefore concluded "we should have a memo in our files reaching the conclusion that a cost savings could be realized by purchases in the quantities for which we propose to offer discounts." A series of internal attempts were made to justify the discounts, some concluding the cost saving to be over $1 a head if twelve carloads (972 head) were ordered instead of one (81 head). Many of their plans had to be reexamined when the Packers and Stockyards Administration started paying more attention to pricing practices and required distributors be given equal treatment. Because IBP's attorneys feared the negative publicity of a Robinson-Patman violation and the possibility of treble-damage antitrust lawsuits too risky, the company terminated discounts until an independent cost study could be done.[32]

In addition to volume discounting, IBP aggressively sought to undercut the price of its rivals. The only way to promote "desirable behavior" among customers, as one IBP vice-president saw it, was to give them a reason based on price. He argued that IBP should reduce prices enough to give retailers an incentive to switch to IBP, or, following the logic of oligopolistic interdependence advanced by some antitrust scholars, "no volume will shift hands and everyone will be less profitable." Smaller packers accused large packers such as IBP with "selling below cost" long enough to drive them out of business and subsidizing the effort with profits from other divisions of the conglomerate. They even asked the Packers and Stockyards Administration to force companies to file reports on individual plants because the aggregated reports failed to detect "unfair competitive practices or acts of monopoly."[33]

During hearings of the House Small Business Committee in 1980 a former executive from IBP came forward with testimony and documents outlining IBP's market strategies. Hughes Bagley, terminated by IBP and employed by Spencer Foods at the time of his testimony, believed "that somebody somehow had to stand up and be counted, or IBP was going to swallow up all of its smaller competition, including my new employer," as part of a "massive takeover by IBP of the packing industry." Bagley testified that IBP became "overly zealous in its attempts to control and monopolize the packing industry" and that "the idea of market domination was discussed continuously by Mr. Holman [co-chairman] and others at IBP. It was almost as if we were waging war against our competitors. It was felt that the best way to achieve market domination was to control the industry at the production level because then we could control the industry at the retail level. Mr. Holman continually stated that the balance of marketing power could be shifted from the retail chains to IBP." Internal IBP documents given to the House committee consistently corroborated Bagley's story.[34]

Cargill and ConAgra also emerged during this period as dominant packing powers. Cargill's entry came with the acquisition in 1978 of MBPXL, the firm that resulted from the merger of Missouri Beef Packers and Kansas Beef Industries in 1974. By 1979 Cargill slaughtered 1.7 million steers and heifers, a year later started a new plant in Dodge City with an annual capacity of 1 million head, and offered boxed beef under the label "Excel Country Cut Beef." Between 1975 and 1980 Cargill and IBP doubled their market share to 30 percent beef slaughter, increased their share of the boxed beef market to 45 percent, and, with Cargill's Dodge City plant and IBP's new Finney County, Kansas plant (1.2 million steers and heifers a year), owned the two largest slaughter plants in the world.[35]

The firm that Cargill outbid for MBPXL was the Omaha-based company ConAgra, which, as the conglomeration trend continued into the 1980s, would be more successful at acquisitions. In 1986 the Supreme Court approved Cargill's acquisition of Spencer Beef (number seven packer in 1978) over the objections of Monfort. Subsequently, ConAgra acquired Spencer,

Monfort, Armour, and Swift. The conglomeration trend even involved IBP when it was acquired by Occidental Petroleum in 1981.[36]

Part of the reason for the success of the new firms stemmed from technology and transportation. Paul Bissell, who worked his way up from the kill floor in 1930 to beef department superintendent in 1956 at Morrell's Ottumwa plant, noted that until 1961, "as far as beef operations were concerned, little had changed back to the 30s." Soon, however, a "second technological revolution" started and new devices like stunners, mechanical knives and hide skinners, power saws, electronic slicing devices, and weighing equipment contributed to a productivity increase of 49 percent from 1960 to 1970. The new boxed-beef process also dramatically reduced costs, since, absent the whole carcass, only one-third as much weight had to be transported. These changes prompted the development of specialized plants that focused on one or two species, not four or five, that involved less capital, less labor, cheaper labor, and more efficient usage of technology, while older firms were stuck with the sunk costs of outdated plants.[37] One economist predicted that such changes reduced the new firms' costs by as much as five to six cents per pound. Because the new packers were primarily located close to cattle supplies, they also enjoyed lower transportation costs, and owing to the emergence of the trucking industry and an improved road system, the old packers no longer possessed the advantage of a large rail network. The use of trucks also made the marketing plans of the new packers more flexible and, since trucks were a relatively cheap capital acquisition compared to the railroad, fostered market entry. Also, the meat-grading system devised by the federal government undermined the older firms' brand-name advantage, reduced the marketing costs of the newer rivals, and increased the buyer's information about quality.[38]

The biggest advantage enjoyed by the emerging firms involved labor costs. The meatpackers who established operations in the rural grain belt, particularly in states such as Iowa, Nebraska, and Kansas, benefited from state right-to-work laws that passed following Taft-Hartley. The United Packinghouse Workers of America thought that such moves purposely sought to take advantage of rural residents who "were not sensitive to traditions and concerns that had prompted the packing community's successful union-building." This labor arrangement contrasted sharply with the pattern bargaining that developed across the meatpacking sector in the 1940s and 1950s, which had taken "wages out of competition." Soon, the older packers were paying about $1.15 an hour more than their newer rivals such as IBP (see table 6).[39]

TABLE 6. PACKING INDUSTRY MASTER AGREEMENT WAGES VERSUS
IBP WAGES (IN DOLLARS PER HOUR)

Date	Master	IBP Slaughter/Processing (boxed beef)
April 1970	3.94	3.34/2.74
October 1973	4.71	3.80/3.19
February 1977	6.47	5.87/5.57
October 1978	7.56	6.22/5.92
October 1980	9.64	8.20/7.90
October 1981	10.69	9.14/8.84

Source: table from Horowitz, "Meatpacking as Paradigm?" 26.

The master contracts also provided for automatic cost-of-living adjustments, which triggered enormous losses during the stagflation of the 1970s and generated resentment among farmers and processors who thought they were wrongfully blamed for high food prices.[40] The near-total organization of meatpacking workers in the 1930s and 1940s had also dramatically increased worker involvement in plant decision making. At the Morrells plant in Ottumwa, for example, workers reached an agreement with management on the speed of the chain, which determined the total number of animals to be killed in an hour. One worker at the Wilson plant in Cedar Rapids recalled that disputes over chain speeds "were probably the main reason for [work] stoppages" during that period.[41]

Another competitive advantage potentially enjoyed by new firms involved the costs of cattle and hogs. During the years of the Big Four, 80 percent of livestock were marketed through terminal markets or local auction houses, where prices were set publicly and rivals knew each others' costs. By 1984, however, only 7 percent were marketed through the terminal markets, while many were vertically contracted by farmers or feedlots and sold directly to the packer. IBP could, for example, underprice rivals by paying less for livestock in regional markets where the competition for supplies was reduced, something they were often accused of doing. Such practices were a precursor to more recent debates within agricultural circles about "captive supplies," where packers contract for the supplies they need and fail to purchase from noncontract farmers in the area.

With the growth of nonunion firms, coupled with the technology that reduced the number of American meatpacking jobs from 274,000 in 1947 to 189,000 in 1972, organized plants suffered enormous losses. Only a few of IBP's plants were unionized; thus, as one IBP official remembered, they could get by with the nonunion plants if a "labor problem" developed. For the most part, however, "management truly ran the plants . . . and no restrictive agreements prevented the company from introducing productivity improvements." IBP's CEO Dale Tintsman admitted bluntly in 1980 that "we're proud of our workers, but basically we can teach anybody to do a job in our plant in 30 days or less—they don't need the skills of an old-time butcher who had to know how to cut up a whole carcass." Such was the strategy invoked in IBP's takeover of the Hygrade plant in Storm Lake, Iowa, in the early 1980s, the company's first large-scale move into the pork business. After buying the plant IBP rehired only a few of the former union Hygrade workers, hired the local union leaders to be part of management, hired enough Mexican and Laotian immigrants to constitute one-third of the work force, and paid one-half the wages the former Hygrade workers received. They were also able to increase chain speeds by up to 80 percent in some plants.[42] IBP's cost advantage grew because union workers were replaced with nonunion workers, mechanization increased efficiency and obviated the need for skilled workers in the plants, and, through the boxed-beef process, the jobs of the specialized butchers of the grocery stores could be eliminated. The competitive pressures exerted by IBP caused tremendous difficulties for unionized plants—California packers, for example, complained that the "substandard wages" of firms such as IBP jeopardized their economic future. Hormel, for another example, with the highest labor costs in the industry at 18.7 percent of sales compared to an average of 10.7 percent for the new packers, struggled through perhaps the most bitter labor dispute of the 1980s at its Austin, Minnesota, plant.[43] Some companies adjusted by following the IBP model. Swift, now a subsidiary of Esmark, shut down three slaughterhouses and paid the workers' severance pay, then spun off its fresh-meat operation in 1981 into SIPCO (Swift Independent Packing Company). SIPCO then rehired the workers at a much reduced pay level. In 1984 Greyhound released its Armour workers, paid their severance pay, then sold Armour to ConAgra, which reopened the seventeen plants with nonunion workers.[44]

By the late 1980s a new set of large firms had emerged in the meatpacking industry. By the 1990s, these firms slaughtered over 80 percent of American beef, a sharp contrast to the much lower concentration levels in the 1960s and 1970s. But the abandonment of Warren-era merger policies by enforcement agencies and the courts—as indicated by the numerous mergers in the meatpacking sector—contributed heavily to the "record volume of food manufacturing acquisitions" in the 1980s.[45] One study concluded that two-thirds of the increase in concentrations levels during the 1980s could be explained by mergers and acquisitions, many of which violated the Department of Justice's own merger guidelines.[46]

More, it seems, could have been done to slow the conglomeration of agribusiness; such steps would have limited the ability to shift costs and profits and improved the farmers' relative bargaining position. Given the sensitive demand for meat, however, the restructuring of the meatpacking industry may have benefited farmers. One economist concluded that packer consolidation "brought efficiencies and saved part of the cattle industry by avoiding the need for still lower prices at the producer level." The market weakness of farmers, it must also be recognized, stems also from their own disorganization and resistance to the cooperation required to defend the fragmented political economy of cattle and hog production rather than solely packer collusion.[47]

WHEAT, CORN, AND BEAN PROCESSING

In the case of marketing livestock, especially after the demise of terminal markets, farmers typically priced their products with individual packers and feeders. In the case of marketing grain and beans, however, farmers typically sold to the local elevator, either private or farmer-owned, which set commodity prices based on a relationship to terminal market prices and futures quotations, the relationship usually based on transportation costs to terminal markets. Problems of competition that did exist tended to be in downstream subsectors of the grain economy such as bakeries in a particular city or manufacturers of cold cereals.[48]

While livestock generates close to half of farm income in some parts of the grain belt, crop production is also very important. Iowa, Illinois, Nebraska, Minnesota, Wisconsin, Missouri, Ohio, Indiana, and Michigan raised around 80 percent of the nation's corn and beans, and the plains of Kansas, Nebraska, and the Dakotas still constitute the nation's most important wheat-producing region. Farmers sold 80 percent of the corn, beans, and wheat they marketed to country elevators, which numbered six thousand in 1959. The market structure of grain processing received even less attention from market analysts than meatpacking, though it did generate antitrust inquiries at times. Because chapter 4 examines grain exporting and chapter 7 examines the federal farm program, the coverage of wheat, corn, and bean processing in this chapter is less detailed.[49]

The competition for the commodities collected by the elevator system comes from several sectors. For wheat, the supply is divided among several hundred flour millers and several exporters, the percentage to each sector determined by supply and demand. The 40 percent of corn production that is marketed (the remainder is used for feed on the farm) is divided among exporters and wet and dry corn processors. Marketed beans are divided between exporters and domestic crushers, who produce meal, flour, oil, and grits.

These sectors showed signs of competition in the postwar period, as indicated by changes in market structure. In 1946, eighty-nine different firms processed 16,000 tons of soybeans; by 1984, thirty-four firms processed 127,000 tons of soybeans. Many of the firms that failed did so because of their older, less efficient, hydraulic presses, which were displaced by new screw presses. Soy oil was also subject to "vigorous competition" from substitute oils and fats but remained a strong industry owing to its many uses, accounting for two-thirds of fats and oils used for margarine and shortening and three-fourths of fats and oils used for cooking and salad oils. Processing by the top four firms was 44 percent in 1946 and grew to 65 percent in 1984, but included different firms. For example, exhibiting the trend toward farmer processing, in 1963 the Kansas-based Farmers Union Cooperative Marketing Association acquired Dannen Mills, which included a soybean-processing plant with a capacity of 750 tons per day. The number of multiplant operations remained steady—thirteen in 1946 and eleven in 1984—but the number of plants within these firms increased. In 1971 ADM possessed 104 soybean-crushing operations, or about 14 percent of the market. The margin between the value of soybeans and meal and oil remained small throughout the postwar period, bringing "low earnings and considerable pressure for structural change." World demand also shaped the industry: prior to the 1970s the U.S. exported 90 percent of the world's soybeans; by the 1980s this dominance was undermined by the large soybean production increases in Brazil and Argentina.[50]

Very little of the annual production of corn makes it to the processing stage. In the 1980s only about 10 percent of corn was not exported or fed to livestock. A good portion of this 10 percent was used in food processing, although some was used for seed, industrial uses, or manufactured animal feed. Wet milling is one subsector of corn processing, which produces starch used in food, paper, textiles, and other items. In the early part of the century mergers created the Corn Products Refining Company, which processed 65 percent of wet corn until it was divested by the government. When World War II ended, the company's share dropped to 45 percent, and by the 1980s a dozen firms processed wet corn, including ADM, Cargill, Corn Products (which changed its name to CPC International), and several farmer cooperatives. Three-fourths of the processing took place in Iowa, Illinois, and Indiana. The overall size of the industry expanded dramatically in the 1970s with the development of high-fructose corn syrup (HFCS), a perfect substitute for beet and cane sugar, and the development of ethanol promotion policies following the oil shocks. The dry corn–milling subsector acquires another 2–5 percent of corn production and produces grits, cornmeal, corn flour, and corn oil. Throughout the century, demand remained a problem: in 1900 over one hundred pounds of dry-milled corn products were consumed per capita, but declined to 7.4 pounds by 1970.[51]

Changes in the market structure of wheat milling also indicate competition. The total number of flour mills declined from 1,243 in 1947 to 361 by 1982. Farmer entry remained possible, however—in 1962 the Farmers Cooperative Commission Company started processing bulgar wheat, over 3 million bushels by 1965, and GTA processed durum wheat. In the 1980s large firms such as General Mills and Pillsbury were bypassed by ADM, ConAgra, and Cargill, which were not even top ten firms in the 1960s. Technology played an important role in market structure changes. In the 1880s Minneapolis developed as the wheat-milling center of the country owing to the growth of higher-quality wheat from the northern plains and the development of a new milling process. In the 1950s Pillsbury developed another process for altering the protein content of wheat, which allowed for the production of different types of flour from the same variety of wheat. Other wheat-milling centers such as Kansas City and Buffalo also developed, putting more pressure on the large Minneapolis-based firms. Falling postwar demand also contributed to the demise of many milling operations. In 1947 war-torn European countries required 100 million tons of flour, but by 1954 the requirement fell to 17 million tons and kept export subsidies at that level. In the 1950s and 1960s domestic consumption of flour grew only about 0.3 percent a year; overall, flour consumption dropped from 10.3 million tons in 1939 to 9.6 millions tons in 1959, and flour consumption per capita reached an all-time low of 110 pounds in the 1970s. The lifting of the wartime wage freeze also increased costs for the smaller mills serving the wartime demand (labor involved nearly 40 percent of the cost of processing). The National Commission on Food Marketing noted that "the period of heaviest mill closings appears to have coincided with the decline in exports and rise in costs." Millers in wheat-growing regions also suffered when it became cheaper to transport wheat than to transport flour (owing to changes in rail rates), triggering a boom in milling near urban areas of consumption. Even though the number of mills dropped by 50 percent from 1950 to 1965, the top-eight concentration ratio dropped from 41 percent to 39 percent. Milling profits seemed modest, accounting for 0.22 percent of sales, according to the estimates of the National Commission on Food Marketing. Because most of the flour is purchased by the baking sector, which possesses enough market power to bargain effectively with millers, the potential for growth of profit margins was minimal.[52]

Food processors also competed for grain with the expanding feed manufacturing industry, which grew from fourteen hundred firms in 1939 to twenty-four hundred by 1959. While a total of 7 million tons of corn were used in both dry and wet milling, 13 million tons were used for feed manufacturing. By 1975 over six thousand feed manufacturing plants existed; entry into the industry remained easy and product differentiation low. In 1982 the top-four firms manufactured only about 20 percent of feed. Farmers could also mix and grind their own feed on the farms.[53]

While competition has kept the grain-processing sector dynamic in the postwar years, substantial concern does exist about the trend toward conglomeration. Cargill, for example, owns plants in wheat, bean, and corn processing, in addition to elevators, exporting facilities, a meatpacking division, and operations in many other sectors. Conglomerates can also shift resources into sectors to finance predatory prices, driving smaller, undiversified firms out the market.[54]

Certain practices have also concerned antitrust officials in the past. In the 1960s, for example, the Federal Trade Commission successfully pursued a suit against the National Macaroni Manufacturers Association, whose eighty-five members manufactured 70 percent of the nation's macaroni. The manufacturers sought to avoid high supply prices by blending durum wheat with hard wheat when durum prices increased, a practice that cost wheat farmers significant profits. To this end, the association met in Minneapolis in 1961 and agreed to send an "expression of opinion" to wheat millers and members of the association about the appropriate blend of durum and hard wheat to be used in macaroni manufacturing. By using blended wheat, the association "ward[ed] off price competition for durum wheat in short supply by lowering total industry demand to the level of available supply." By fixing the composition of their most important raw material, the FTC concluded, macaroni manufacturers substantially affected the price of durum wheat. The Seventh Circuit agreed, recognizing the ability of monopsonistic buyers in the agribusiness sector to undermine farmers economically.[55]

The anticompetitive trends of these changes are also constrained by the marketing options of farmers, their tendency to integrate forward into processing, and their ability to store grain. Since at least the nineteenth century farmers have at times processed their own products, a trend that has continued into the late twentieth century. Unlike livestock, which require marketing at certain times, crops can be stored and therefore offer farmers more marketing flexibility. In the postwar period farmers' storage capacity increased dramatically—by 1982, 60 percent of the nation's total grain storage capacity was located on farms. The grain- and bean-processing sector is also shaped by the international market for commodities—where a very large percentage of grains are sold—and by the workings of the government farm program, subjects that are covered in more detail in chapters 4 and 7. In the early 1960s, for example, 60 percent of the wheat farmers sold was traded internationally, and therefore the price fluctuated based on international demand-and-supply conditions, at times triggering charges about international grain traders similar to those leveled at domestic meatpackers.[56]

Broad indicators of horizontal competition between meatpackers and grain processors do not always translate into competition for the supplies of farmers. As a result, farmers and policy makers have emphasized the importance of enhancing the ability of farmers to bargain effectively with agribusiness buyers, a process examined in the second half of this book.

4

THE GRAIN-TRADING "CARTEL"

> The five companies maintain a strangle hold over the world's grain supply and constitute a food cartel unprecedented in world history.
> ROGER BURBACH, *THE PROGRESSIVE*

> Just control Cargill. FARMER OPINION

Ben Hogan's turkey farm was in jeopardy when feed costs increased more than a third owing to a "huge international grain trade agreement overseas." The feed cost increases could have come after the massive failure of Soviet harvests in the early 1970s, which created a demand that was met with a large expansion of American grain exports. The size of the Russian grain deal, coupled with allegations of profiteering on the part of grain companies and secret deals on the part of the Nixon administration, triggered much talk about a "grain cartel" or a "grain oligopoly." In the mid-1970s Senator Dick Clark (D, IA) argued that "our exports are controlled by a very few multinational firms," which "almost totally dominate our markets and the world's markets," and the Food Action Committee attacked the "monopolistic grain corporations."[1]

The talk resonated with those who feared corporate farming and processor collusion and viewed farmers as pawns in superpower politics, especially after allegations of insider information on the part of the grain companies' sales of 1972, the subsequent Ford and Carter grain embargoes of 1974, 1975, and 1980, and widely publicized proposals for the United States to begin wielding its "agripower" in international relations. Many critics assumed that a tightly organized, collusive cartel of grain companies existed, as indicated by an exchange between Senator Clark and Verl Loyland, who operated the cooperative Finley Farmers Grain & Elevator Company.

Clark: You seem to feel there is collusion between these companies, that they get together and decide what they are going to offer?
Loyland: We definitely feel so.[2]

The assumption remains doubtful given the constant problems with different schemes to rationalize world trade, American use of the P.L. 480 program, trade restrictions and state trading by other nations, the cold war geopolitics that constantly invited government interference in the grain trade, and the ease of entry into export markets. Like meatpacking, then, the grain trade displayed signs of competitiveness, as best evidenced by the exporting activities of farmer organizations such as the NFO and various cooperatives.

THE FAILURE TO RATIONALIZE WORLD AGRICULTURAL TRADE

The difficulties of collusion in grain export markets are underscored by the many failed attempts of national governments to rationalize world agricultural trade. In the 1920s, for example, in response to the long-simmering demands of agricultural groups for tariff protection equivalent to the protection afforded industry, and in immediate response to the post–World War I farm crisis, the Republican administrations extended trade protections. Farmers were angry about the double standard applied to tariff policy—while industrial America enjoyed high tariffs offering protection from foreign competition, agricultural commodities and thus farmers were subject to the whims of the international market. The demands for protection complimented the Republican administrations' efforts to organize agricultural marketing and prevent price destabilization from cheaper foreign commodities.[3]

Just as Secretary of Commerce and later President Hoover attempted to coordinate and stabilize the domestic agricultural system and protect it from disruption through high tariffs, he also attempted to coordinate and stabilize global commodity markets. Hoover and others viewed the sugar market, for example, as disastrously unstable and therefore in need of progressive government coordination of the variety Hoover espoused. The American market included more than just continental producers of cane and beet sugar—it also included Cuba (which had a tariff exemption after 1902), Hawaii, Puerto Rico, and the Philippines, which were American territories. After World War I, similar to many industries, sugar was seriously maladjusted: in 1923 Cuban sugar sold for well over four cents a pound on the New York market, but by 1929 the price dropped to less than two cents. As a result, the Hoover administration encouraged the sugar producers to invoke his associational plans for a self-governing, cooperative arrangement to reduce production and boost prices. Just like Hoover's domestic stabilization programs, the sugar effort ultimately failed, and for most of the early years of the 1930s sugar sold at around one cent a pound.[4]

Although Hoover sought to balance and order markets through voluntary cooperative action encouraged by the state, he generally wanted market forces to shape economic activity—he believed a largely free economy to be necessary for continued American economic growth and prosperity. In the case of the global rubber market, Hoover believed that the American economy suffered because of an absence of competition. Hence he promoted State Department efforts to coordinate the investment of American corporations in overseas rubber plantations in order the break the British-sponsored Stevenson rubber cartel. As in the case of sugar, the rubber cartel was organized in response to a price plunge from sixty cents a pound in 1918 to twelve and one-half cents by 1922. Hoover saw the rubber cartel, however, not as a stabilizing or rationalizing effort but as an effort to gouge or "super-charge" the American consumer. When rubber prices surged to $1.20 in 1925 Hoover whipped up anti-foreign monopoly sentiment, created support for private-sector buying pools to counter the cartel, and promoted business efforts to create alternative sources of rubber. He mounted similar campaigns against foreign cartels in coffee, oil, and potash. The failure to boost sugar prices and Hoover's success at undermining the rubber arrangement underscore the problems inherent in maintaining cartel-like arrangements in international agricultural commodities throughout the twentieth century.[5]

In addition to global sugar and rubber regimes, and perhaps his greatest interest and area of concentration, Hoover applied his economic ideas to the domestic organization of core agricultural commodities such as wheat. As Hoover saw it, agriculture depended too heavily on unstable and unreliable foreign markets and should therefore attempt to isolate itself, taking advantage of America's "remarkably self-contained" economy, and concentrate on coordinating domestic production to domestic demand. The voluntary, federated agricultural cooperatives that he envisioned would manage this process failed to coordinate production properly, setting the stage for mandatory acreage reductions under the New Deal's AAA.[6]

Hoover's unwillingness to use state power obviated one of the decade's largest efforts by farmers to boost prices artificially. The McNary-Haugen bill, twice vetoed by President Coolidge and vehemently opposed by Hoover, would have forced the federal government to establish two prices for agricultural commodities, a high domestic price and a price low enough to allow "dumping" on the world market—a policy that would have been resented by foreign governments trying to protect their farmers, a fact quickly recognized in the post–World War II period. Hoover's unwillingness to engage in state-sponsored dumping, his inability to stabilize global commodity prices (for rubber, wheat, sugar, and coffee, for instance), and the failure of his cooperative plan to raise domestic agricultural prices contributed heavily to his electoral defeat in 1932 and, unwittingly, to the construction of the postwar agricultural order.

In response to the agricultural crisis of the 1930s, and in contrast to Hoover's failed voluntary cooperative efforts, Roosevelt's Agricultural Adjustment Act of 1933 designated a national allotment of key commodities, paid farmers to produce within the limits of the allotment, and thereby increased prices through artificial scarcity (see chapter 7). To prevent the price support system from being overwhelmed, the government built in protections from foreign competition (the secretary of agriculture was responsible for determining if imports would undermine the farm program), angering groups such as Cuban sugar and Canadian wheat producers. Specifically, Section 22 of the AAA authorized import quotas or fees, and Section 32 authorized export subsidies on wheat, which were used continuously from the end of the Korean War until the Russian grain sales of the 1970s. The agricultural program conflicted with the diplomatic efforts of the Department of State, as indicated by the bitter clashes between Secretary of State Cordell Hull and AAA director George Peek over the Reciprocal Trade Agreements Act—a rivalry continued in the postwar years by Secretaries Kissinger and Butz, for example.[7] Hull would win the argument over the bill, but agriculture's defenders would insure exemptions protecting agriculture from foreign competition, both in the 1930s and after World War II.[8]

During the early years of the New Deal, representatives of the Roosevelt administration also took part in an economic conference in London that addressed the idea of government-sponsored international commodity agreements to establish minimum and maximum world prices for products—agreements some would call cartels (farmers, on the other hand, viewed the grain companies as the "cartel"). The International Wheat Agreement (IWA) of 1933 involved nine exporting nations and thirteen importing: the exporting nations were to reduce acreages 15 percent, and importing countries agreed not to increase their acreages and lower tariffs in order to stabilize prices. The agreement quickly fell apart when Argentina broke its quota after a bumper crop, but the remnants of the program helped shape the postwar international agricultural trade.[9] The fragments of earlier farm policies—the domestic production control and price support policies of the AAA (which filled the vacuum left by the collapse of Hoover's farm program), the attempt at international cooperation to coordinate and stabilize the global wheat trade, and agriculture's exemptions from increasingly stringent rules governing world trade—all combined after World War II to shape the structure of the international agricultural order.

After the war many world leaders considered the economic nationalism associated with efforts to raise prices during the interwar years as partly responsible for the global war—as Hull argued, "if goods can't cross borders, soldiers will." The trade negotiations that ensued in Havana in 1947 led to an agreement to reduce world tariff levels and to establish an International Trade Organization (ITO) to police trade disputes and the General Agreement on Tariffs and Trade (GATT), all of which led to tariff concessions on two-thirds of the world's trade. Due to the influence of agricultural interests, however, agriculture was exempt from the GATT rules banning import quotas. The ITO, which had been seriously jeopardized from the beginning owing to the issue of American agricultural subsidies, ultimately failed to gain approval from the U.S. Senate. Because of cold war priorities, the United States also agreed to the creation of regional trading blocs, which undermined the trade reciprocity principles of the GATT and justified the creation of arrangements such as the EEC's Common Agricultural Policy.[10]

Despite these failures, or perhaps because of them, in 1947–48 negotiators again attempted to stabilize world agricultural prices and prevent "ruinous competition" through another International Wheat Agreement. Objections to "state trading" were raised by the National Grain Trade Council, the Millers' National Federation, the Flour Millers' Export Association, firms such as Pillsbury and General Mills, and the Farm Commissioners' Council, made up of the farm commissioners from individual states. Nevertheless, the IWA negotiated became effective in 1949 and lasted for four years. The exporters in the agreement (Australia, France, Canada, and the United States) assured a supply of grain to importers (some forty countries) in exchange for assurance that the importers would buy a supply of grain—the amount to be exchanged was fixed according to an annual quota schedule. The agreement established a price range above which the exporters could not sell and below which the importers could not buy on the amount fixed

TABLE 7. INTERNATIONAL WHEAT AGREEMENT, 1953 (IN BUSHELS)
MAXIMUM PRICE: $2.05
MINIMUM PRICE: $1.55
U.S. SUPPORT PRICE: $2.49
U.S. DOMESTIC PRICE: $2.27
WORLD PRICE: $1.86

Exporters' "guaranteed quantities"		Importers' "guaranteed quantities"	
United States	196.5	Germany	55.1
Canada	153.1	India	36.7
Australia	45.0	Japan	36.7
France	0.4	Netherlands	28.5
		Belgium	23.9
		Egypt	14.7
		South Africa	13.2
		Brazil	13.2
		+ 36 other countries	175.0
Total	395.0	Total	395.0

Source: Department of Agriculture Proposal in U.S. Participation in International Wheat Agreement, Oct. 5, 1955, FF U.S. Participation in IWA, DB 5, U.S. Council on Foreign Economic Policy, Policy Paper Series, Eisenhower Library.

by the quota. Since the annual quota accounted for only 60 percent of the world's wheat trade (37 percent of American exports in total), however, prices continued to fluctuate. Also undermining the workability of the agreement was the refusal of Argentina and the Soviet Union to join in 1949, the dropping out of the world's largest importer, Great Britain, after the negotiation of a new agreement in 1953 (because they thought the new price maximum of $2.05 too high), and, unlike the 1933 agreement, the absence of production controls. One Minneapolis trader argued that under the IWA "Uncle Sam would be acting the part of a simpleton. The United States would agree, for a five year period, to be a sucker." The trader feared that when demand increased, Argentina and the Soviet Union could charge whatever price they could receive, whereas the American commodities organized under the IWA would be subject to the price maximums. An official at Continental Grain Company argued that the agreement precluded the United States from beating out other competitors for import markets based on more efficient American production techniques. Proponents such as Senator Arthur Capper of wheat-rich Kansas countered that the IWA would insure at least some foreign markets for American commodities and an outlet for government stocks resulting from the farm program that were expensive to store.[11]

The IWA continued to suffer in the late 1950s as the United States hesitated to continue participating in an arrangement that seemed of little help to the country. Officials recalled, for example, that the price surge associated with the Korean War cost the United States money owing to the agreement's price ceilings. But in the summer of 1958, with 800 million bushels of wheat in storage and another 1.2 billion expected to be harvested in the fall (when only 600 million was consumed annually), the Eisenhower administration consented to another IWA, believing that it at least moved American surpluses into international markets, that wheat traded between the agreement's price limits moderated overall price swings, that it would reduce the cost of export subsidies, and that nonparticipation "could have an adverse effect on foreign relations." The resulting 1959 IWA was weakened, however, when it dropped the arrangement that importers buy a fixed quantity of wheat and instead buy a quantity of wheat based on a percentage of their total imports, resulting in the importers' use of substitutes and greater self-sufficiency. In 1962 language was inserted into the IWA ending the requirement that domestic agricultural policy be made with the integrity of the IWA in mind, giving exporters and importers "complete liberty of action" domestically (a far cry from the production controls of the 1933 IWA).[12]

In the late 1960s, with farm surpluses continuing, policy makers attempted another international agreement to stabilize and coordinate the wheat market, in spite of the problems associated with previous efforts. The result was the International Grains Agreement negotiated in 1967 (the term *grains* was used instead of wheat because prices were pegged to several different varieties of wheat instead of just one prominent variety, Manitoba No. 1). By the time the agreement came into effect in 1968, however, the pricing structure agreed to had already been overwhelmed by that season's large harvests. Unable to unload their massive surpluses at the minimum price established under the IGA, major exporters such as Canada and France willfully violated the agreement, and the United States started to argue that the minimum prices agreed to were only "guidelines." In 1971 the Wheat Trade Convention reconvened, but no purchase or supply obligations or pricing structure could be agreed to. The agreement lapsed despite the support of the NFU, NFO, Grange, Grain Terminal Association, National Wheat Growers, National Soybean Growers, National Corn Growers, the National Rural Coalition, and the Agribusiness Accountability Project, and the warning of the Kansas Farmers Union that the absence of an agreement would exacerbate the "chronic rural depression [that] has already brought calamity in our cities and made wasteland of much of rural America." An attempt to negotiate another agreement in 1978 also failed. The extreme difficulties involved in coordinating international commodity markets even with the aid of governments underscore the difficulties of collusion among private firms.[13]

P.L. 480

In the early 1950s, with only a limited amount of sales organized under the IWA, Marshall Plan sales to Europe shrinking, the Korean War ending, and USDA commodities piling up, political leaders were still searching for the best method to manage the agricultural surplus.[14] With the election of Dwight D. Eisenhower in 1952 the emphasis of previous decades shifted— Ezra Taft Benson's promotion of the "freedom to farm" included more emphasis on exporting. Economics and politics, however, undermined Benson's ideological policies, as economic distress in agricultural areas prompted congressional opposition to cutting support prices. The export promotion policy caught on, however, and essentially turned into a form of state-sponsored trading, further undermining the potential for maintaining a private grain cartel.

The most important attempt to increase exports stemmed from an emerging movement to scale down American surpluses through government-aided sales and donations to other countries. Although philosophically opposed in the early stages, and heeding Secretary of State John Foster Dulles's warnings about opposition from friendly nations, President Eisenhower changed his position. Part of the reason for the change of heart was his frustration with bureaucratic complications that hindered his efforts to supply aid to Pakistan during a devastating drought; moreover, he realized that congressional momentum was too powerful to thwart. As Trudy Peterson described it, "Congress . . . had now seized on the surplus-disposal-cum-relief idea, and the administration decided that it could 'live with it.'"[15]

Eisenhower set the USDA, the Commission on Foreign Economic Policy, and an interdepartmental committee to work, and the policy resulting from the USDA report was announced during the president's farm message in January 1954. The proposal set aside $2.5 billion in Commodity Credit Corporation commodities (the CCC managed the surplus) for disaster relief, foreign aid, domestic school lunch programs, and reserves. The interdepartmental committee also sought to co-opt a Senate bill passed during the summer of 1953 authorizing $500 million in foreign currency sales.

The final administration proposal increased the sum to $1 billion and was included in the $2.5 billion set-aside announced by the president in January. The final administration bill competed against sixty other surplus disposal bills introduced in the session and was signed into law in 1954.[16]

The resulting legislation, known as the Agricultural Trade Development Act of 1954, or simply Public Law 480, significantly altered the nation's ability to manage its commodity surplus problem. The administration was now able to negotiate the sale of $700 million in surplus CCC stocks for foreign currency in the next three years and funnel the money back to the purchasing country for economic development. In addition, $300 million more could be disposed of via contributions to humanitarian or famine relief efforts. The program became largely a "free gift" to needy countries that would otherwise have to spend their foreign exchange for food or "do without." At the same time, American farmers partially avoided the competitive prices of foreign trade and the need to find foreign exchange, and as a result, exports grew from $3 billion in 1953 to $5 billion in 1960. By the late 1950s one-third of American economic aid took the form of these local currency sales, and 70 percent of wheat exports would be handled by the program; by 1962 one-half of soybean oil exports were moved under P.L. 480. Presaging the "agripower" debates of the 1970s, and a reflection of the growth and power of the program, were debates over the control of the program in the 1960s. Farm state legislators such as George McGovern and Walter Mondale fought to keep the program in the USDA, whereas others wanted to move into the State Department.[17]

The absence of international rules governing agricultural trade and surplus problems in other countries fostered the existence of P.L. 480-type programs and state trading in other countries, constituting another barrier to the formation of a private grain-exporting cartel. All Canadian grain exports, for example, were sold by the Canadian Wheat Board, whose mandate is "to market as much grain as possible at the best price that can be obtained," and Australian grain producers marketed their commodities through the Australian Wheat Board. An example of the clash between P.L. 480 and another state-trader's efforts involves Iran. In latter part of the 1960s, 95 percent of soybean oil exports were subsidized by P.L. 480; an important export market was Iran, which seemed at risk when the Soviet Union began dumping sunflower oil. More P.L. 480 credit stopped the Soviet threat. The same strategy prevented Australian and Canadian wheat from displacing American commodities. Such activity provided a difficult atmosphere for private collusion.[18]

THE COMMON AGRICULTURAL POLICY (CAP)

In the 1960s fears of "neomercantilism" were fueled by the emerging power of the European Economic Community and Great Britain's decision to seek membership in 1961. President Kennedy and other officials decided to seek new executive authority to negotiate reductions in tariff levels in hopes of fully participating in the British-EEC talks; agriculture became a prime concern in these negotiations. The Common Market's Common Agricultural Policy (CAP), anchored by the agreement between France and West Germany to lower trade barriers for the benefit of German industrial goods and French agricultural commodities, threatened to close the United States out of one of the world's largest markets for farm goods. Originally a large importer in the postwar years, by the latter part of the 1960s European farm output increased at three times the rate of population growth, forcing the Europeans to find export markets and compete with U.S. commodities. Since Kennedy believed that American agricultural exports could help fulfill his hopes for expanded economic growth, in addition to relieving the widening balance-of-payments deficit and reducing the costs of domestic agricultural programs, he put Secretary of Agriculture Orville Freeman in charge of leading an agricultural export drive. Knowing that the six members of the EEC consumed one-third of American agricultural exports (Germany, Italy, Belgium, Holland, Luxembourg, and France), Kennedy asked the EEC to consider foreign interests before deeply entrenching the CAP to the exclusion of American farm imports. He was aided by the Cargill official William Pearce, then president of the Feed Grains Council, who helped coordinate the attack on the CAP.[19]

In the ensuing Dillon Round of the GATT, the EEC made an offer to reduce all tariffs 20 percent across the board. The representatives of the Kennedy administration were hampered in their negotiating efforts, however, by the limited authority to bargain on trade issues conferred on them by Congress—they were still operating under the system for trade negotiations established under the New Deal's Reciprocal Trade Agreements Act.[20] Although this limited the administration's flexibility, and the round ended with an agreement to reduce tariff levels a mere 10 percent, the Kennedy administration was able to secure a promise from the EEC to maintain current levels of American agricultural imports. And as a result of the weak American bargaining position, Kennedy was able to gain momentum for his efforts to expand the administration's authority to negotiate. The diplomacy of agricultural trade did not simply revolve around the interests of American farmers and the bureaucratic apparatus charged with negotiating trade deals. It also involved the stability of the Western alliance, American leadership in the alliance, and as George Ball argued, the question of whether the Third World remained "closer to Lincoln than Lenin." With these justifications, Kennedy promised Congress he would seek a trade bill in 1962 and promised to make it his top legislative priority of the year. The resulting passage of the Trade Expansion Act (TEA) of 1962 granted the president the authority to lower tariffs by up to 50 percent.[21]

After making no progress on the agricultural front during the Dillon Round, the beginning of the post-TEA Kennedy Round meant Secretary Freeman would spend much energy seeking more openings for American farmers. If European tariffs could be cut 50 percent by 1970, the secretary's advisors told him, American livestock exports could increase by 34 percent and crop exports by 21 percent over 1961 levels. Instead, Freeman feared that EEC agricultural policy would result in privileged status for French wheat, Italian rice, and Danish dairy products, all to the exclusion of American commodities. President Kennedy fully supported his secretary's efforts and even threatened to roll back American troop levels in Europe and Third World aid if progress was not made. From 1959 to 1968, the time period in which the CAP was organized, European protection levels rose dramatically: for meat, from 19 to 52 percent; for dairy, from 19 to 137 percent (from 30 to 350 percent for butter); for cereals, from 14 to 72 percent. While the Netherlands levied a fee of fourteen dollars a ton on U.S. flour before the CAP, afterward the levy jumped to forty-five dollars a ton. Secretary Freeman told the Grain Terminal Association that "a variable levy is simply a device for preventing any imports from coming in below domestic support prices." From 1966 to 1969 American agricultural exports to the EEC subject to the levies of the CAP tumbled nearly 50 percent. According to the *Farm Journal*, by the end of the Kennedy Round, which lasted from 1964 to 1967, "the much-advertised fight that the administration was going to put up for U.S. farmers ended in almost complete capitulation." The United States also failed to liberalize agricultural trade by maintaining the restrictive arrangements of the U.S. Sugar Act, imposing more restrictions on dairy imports, passing the Meat Import Act, and continuing its own price support programs.[22]

The potential leverage of American negotiators on the agricultural issue included a full-fledged trade war. Although CAP commodities favored by the variable import levy system included wheat, coarse grains, and pork and were all contentious trade issues, perhaps the best example of the U.S.-EEC commercial rivalry was the "chicken war." From 1929 to 1961 the output of American chickens grew fiftyfold, and from 1948 to 1964 prices dropped from thirty-five to fourteen cents per pound; as a result, poultry exports surged. In the case of Europe, American exports of poultry, totaling only $3.6 million in 1958, grew to $53.5 million by 1962. The growth of poultry exports was aided by title 1 of Eisenhower's Food for Peace (P.L. 480) program, under which the West Germans used marks to pay for 4 million pounds of poultry in 1956. The booming American exports, however, fell victim to the organization of the CAP in 1962, specifically regulation 22, which outlined plans for a common market in poultry meat. By 1963 American poultry exports had been cut nearly in half. The Institute of the American Poultry Industries and its members began to lobby the Kennedy administration and powerful southern legislators with poultry-dependent constituencies for help. The crisis became so acute that President Kennedy wondered whether the "Grand Alliance [was] going to founder on chickens."

In May 1962 representatives of the International Federation of Agricultural Producers met in Washington with Kennedy and urged him to work against the impending EEC restriction on American poultry. The American negotiating delegation prepared memorandums and position papers and vigorously sought to head off the European threat; Secretary Freeman made clear to the Europeans that Kennedy had the authority to retaliate if no progress was made. American poultry producers also began to lobby Congress for retaliatory measures against European wine, cheese, and cars. Many in Congress were listening; in February 1963 Senator Harry Byrd (D, WV) held a hearing on the chicken war, and twenty-three senators and thirty-one representatives attended. The question of competing interests between foreign policy and agricultural policy was front and center as some witnesses accused the administration of pulling its trade punches in favor of the stability of the Western alliance. Senator Wayne Morse (D, OR) argued that it was time "for a complete review of our foreign policy toward Europe, and that means we have got to look at NATO now and make it perfectly clear that they need us more than we need them." The news commentator Drew Pearson argued that economic retaliation was "not the way to build Western unity against Communism." The administration would have to make a choice between the domestic agricultural interests, with their stake in the construction of a profitable political economy for agriculture, and the foreign policy experts, with their stake in international diplomatic efforts to contain communism. It was an open question whether agriculture would win, as it had with domestic subsidy programs in the 1930s and exemptions from free trade rules during the 1930s and 1940s. In the case of chickens, unlike the P.L. 480 program, agricultural and geopolitical ends could not be served simultaneously.

The Kennedy administration took its case to Geneva and the GATT and promised, in the case of failed negotiations, to seek "balancing compensation elsewhere in our trade exchanges." Again, the *New York Times* wondered whether the EEC was on the brink of becoming an "inward-looking, high-tariff club" and argued that the answer would determine the "future shape and even the fate of the Atlantic Community." C. L. Sulzberger thought that if the United States was denied access to the EEC market "the Whole Grand Design for NATO defense, interdependence and a tightening Western comity of nations could scatter like feathers in a hen-house." The U.S. Senate took time to pass a resolution asking the American delegation in Geneva to maintain access to American export markets. The negotiations started with an American offer to participate in a nonobligatory arbitration conducted by a GATT panel. The panel concluded that the EEC had violated GATT rules with its restrictive policies, but it also ruled that the economic damages caused the United States were substantially less than that nation claimed. The United States and the EEC accepted the conclusions of the panel and sought to implement the recommendations, averting a full-fledged trade war. The episode underscores the complexity of international agricultural trade and the level of state interference and cold war concerns involved in international agricultural trade. Secretary Freeman's economic advisor, Willard Cochrane, later regretted fighting the "chicken war," since "in every country [he] visited from Spain to Malaysia the agricultural leaders in those countries . . . proudly show[ed him] their developing broiler industry" and "new broiler factories," and he was sure "many, many more [were] going to be built around the world in the next few years, including Western Europe," owing to the transferable technology and the efficiency with which broilers converted feed into meat.[23]

In the early 1980s another GATT panel approved EC flour export subsidies, and the U.S. deputy special trade representative commented that the decision "introduces the law of the jungle" into the agricultural trade. The United states then subsidized a large sale of flour to Egypt, which triggered angry EEC condemnations of the "brutal [American] takeover" of the Egyptian market. German foreign minister Hans Dietrich Genscher informed Secretary of State George Schultz that the sale could damage U.S.-EEC relations; British foreign minister Douglas Hurd worked hard to "keep the arguments on agricultural trade under control so they don't produce damaging political results"; a French trade official concluded that "we are on the verge of war"; the *New York Times* argued that the "competition for the [grain] trade has become ferocious." The consequences of the CAP were well summarized by the longtime University of Chicago analyst of farm markets D. Gale Johnson: "The Common Market, instead of being an importer, is pushing all the grain it can out of [the] backdoor. None of it is good for the American farmer." American policy makers responded to the "EC's offensive" in 1985 with an "offensive weapon" known as the Export Enhancement Program (EEP), a subsidy system designed to regain American market share in grain exports. Senate Majority Leader Robert Dole persuaded the Reagan administration to accept EEP in exchange for farm state senators' votes on the Reagan budget over the objections of Cargill, other grain traders, and consumer groups. The "U.S.-EC subsidy war," according to the president of the North American Export Grain Association, was waged in part to improve the American bargaining position in the GATT talks surrounding the issue of agriculture—squabbles over agriculture still stretched the Uruguay Round of the GATT to eight years. The EEP, according to Trade Representative Carla Hills, maintained "the credible threat of retaliation." In sum, the EC created enormous problems for American grain exporters: in the 1960s the EC imported 15 million tons of grain, but by the early 1980s, with the help of the CAP and export subsidies, they were exporting 12 million tons.[24]

THE SOVIET SALES

The pitfalls and complexity of international agricultural trade were also underscored during the 1972 American grain sales to the Soviet Union. The massive grain purchases stemmed, in part, from the evolving sophistication of Russian eating habits—the five-year plan then in effect promised to boost meat consumption as much as 25 percent—and, more specifically, a 1970 Polish riot protesting an increase in food prices, the poor quality of food, and a shortage of red meat. To make up for the shortage caused by poor harvests, the Soviets opted for large-scale imports, as opposed to reducing domestic consumption and risking more social rebellion. A similar situation in 1963 and 1964 led the Soviets to the American granary, but the purchases were not ongoing. Many opposed the 1960s deal, including all but one member of the Iowa congressional delegation, members from grain-dependent states such as then-Congressman Dole of Kansas and Senator Mundt of South Dakota, and former vice-president and cold warrior Richard Nixon. Proponents like George McGovern, however, believed the exports would bolster farm prices, deter the Soviets from spending the grain dollars on weapons, and improve superpower relations.[25]

In the early 1970s, the politics of Soviet grain purchases changed. The nation's relative global economic position slipped considerably in the intervening decade, and American economic largesse dwindled. As OMB director George Schultz declared, "Santa Claus is dead." Now-President Nixon and his advisors thought increased grain sales would strengthen the dollar, reduce the mounting balance-of-payments deficit, address the trade deficit (in 1971 the United States ran the first trade deficit since 1888), pare down American commodity surpluses, and improve superpower relations. He therefore agreed in 1971 to end the licensing requirement for grain exports to Russia and China and the requirement that 50 percent of all shipments be carried on American merchant marine vessels (a Kennedy concession to get the maritime unions to load the grain ships in the early 1960s). Given the opportunity, and assured of a government export subsidy guaranteeing sixty-dollars-a-ton gateway prices for wheat (they feared prices would rise suddenly and prevent them from supplying the Russians profitably), the grain companies jumped. In the summer of 1972 the companies funneled almost 12 million tons of grain to the Soviet Union. Inflation-conscious consumers, who were ultimately forced to pay higher food prices owing to increased demand for grain, denounced the deal. Some farmers who were unable to take advantage of the higher prices because of the haste, secrecy, and what presidential candidate George McGovern called the "blatant special interest favoritism" of the deal, also objected. The sale ushered in a period of food shortages that would end American export subsidies for wheat and even prompt government intervention to embargo the exportation of soybeans, the only government action to impede American agricultural exports since World War II.[26]

The growing importance of the overseas grain sales underscored longstanding fears among farmers about the power and growing concentration of the grain trade. Although the industry involved some thirty firms in the 1920s, in the 1970s it involved six companies exporting 96 percent of wheat, 95 percent of corn, 90 percent of oats, and 80 percent of sorghum, a fact highlighted publicly during the Russian grain sales. The sales were part of a larger export boom from 1971 to 1975 that equaled the total growth in the international grain trade for the entire postwar period. The changes meant greater dependence on the international market and, as many farmers saw it, greater dependence on a global cartel of grain-trading companies that unfairly manipulated prices and were not subject to any effective control. Farm groups were also angered by the grain trade's opposition to international commodity agreements like the International Wheat Agreement and the companies' interest in keeping prices low to remain competitive.[27] The skepticism about the grain companies' involvement in sales to Russia deepened when it seemed that the cold war geopolitics practiced by the "striped pants" at the State Department outweighed the interests of the farmers and the integrity of the political economy of agriculture—Senator Frank Church (D, ID) even had to ask President Ford to include the secretary of agriculture in U.S.-Soviet grain negotiations. Earl Butz later recalled that "my fiercest battles were with Henry Kissinger. He was always trying to get his hands on food for a foreign policy tool."[28]

The smaller supply of food in the early 1970s, not only in the Soviet Union and the United States (after the large grain sales) but throughout the world, substantially increased the value and power of the productive capacity of American agriculture. After decades of attempting to dispose of burdensome stockpiles of American abundance, attempts that often had to be limited owing to the protests of allies, talk grew of the United States' wielding "agripower" as a diplomatic weapon. With 75 percent of the world's grain exports as a diplomatic tool, President Nixon in 1972 secured the SALT treaty, which, according to one author, directly hinged on American grain sales.[29] In 1972 and 1973, 70 percent of P.L. 480 food went to aid the pro-American governments in South Vietnam and Cambodia. A food crisis in Chile, exacerbated by a cutoff of American P.L. 480 aid, suspended grain sales and credit, and American manipulation of international lending, among other efforts to "destabilize" the government, helped bring down the Allende regime in 1973. Kissinger used access to American grain as a tool for brokering Middle Eastern peace, and as a result, by 1978 Egypt was the largest recipient of P.L. 480 aid. In 1975 the United States also sought to use its "grain power" to convince the Soviet Union to exchange 10 million tons of oil at below-OPEC prices for access to American grain stores.[30]

The OPEC oil shocks, coupled with the economic instability associated with the collapse of the Bretton Woods exchange system, stagflation in the United States, and a greater awareness of environmental damage and limits to economic growth, opened the door to what some feared would become a global resource war in the 1970s. In 1974 the National Strategy Information Center released *Can We Avert Economic Warfare in Raw Materials? U.S. Agriculture as a Blue Chip*, a study exploring the ways farm exports could be used as an economic lever. A later, expanded version of the study argued that the United States was in a prime position to extract concessions from foreign powers given its massive capacity to produce agricultural commodities and the limits on other countries' capacities. The study pointed to the Soviet Union as being especially vulnerable to the use of American food power. After the foreign policy setbacks of the 1970s, the CIA thought with the use of food power the "United States might regain the primacy in world affairs it held in the immediate postwar period."[31]

To the extent that American agriculture became a pawn of international resource politics and cold war diplomacy—at a time when the surpluses had become a blessing and farm prices were rising—farmers objected. President Ford invoked grain embargoes in 1974 and 1975 to protest Soviet adventurism in the Third World and to help stabilize domestic food prices, watching closely the correlation between Russian purchases and increases in the consumer price index. Bitter protest from farmers forced him to recant during the 1976 presidential election: "We will never use the bounty of America's farmers as a pawn in international diplomacy. No Embargoes." Ford promised to prevent a return to the days of surpluses: "We must sell grain, not pile it up in storage." Despite the promise, the anger in agricultural regions remained intense: Governor Robert Ray of Iowa objected to the "meddling" in international trade that created "havoc in the market place"; the NFO estimated the financial loss to farmers of the 1974 embargo at $10 billion; Ford lost the May 1976 Nebraska primary to Governor Ronald Reagan partly because of the embargo/export control issue. The anger of farmers explains Ford's decision to choose Senator Dole of Kansas as his running mate in 1976, since he had fought the embargo in the Congress. Governor Jimmy Carter also pledged not to embargo food in the 1976 campaign but broke the pledge after the Soviet invasion of Afghanistan in 1979, triggering a "tremendous backlash from farm interests." In the election of 1980 Reagan promised to reverse Carter's embargo policy. Soon after Reagan won the election, however, he came under intense lobbying from Secretary of State–designate Alexander Haig, who insisted that the embargo remain in place and asked for personal authority over American food policy so he could use it as a diplomatic weapon against the Soviets. Reagan refused, and a few months into his first administration he ended the embargo, declaring that the "granary door is open."[32]

On top of foreign policy interfering in the grain trade, and undermining any would-be cartel, were the information problems involved in the Russian sales, a critical ingredient in collusive behavior. When the Soviets approached the U.S. government about securing CCC-financed grain, American officials did not know the extent of Soviet demands (they didn't believe their own embassy's crop reports) and that the Soviets were also negotiating with Continental Grain Company for private cash sales. After they reached an agreement with government officials on the credit issue, the Soviets quickly negotiated several more contracts with other grain traders before the extent of their purchases became known and thereby enjoyed a lower price. The grain companies were so secretive that they didn't know the Soviets were bargaining with all of them. If they had known the size of the Soviet request they could have received a better price; Henry Kissinger admitted that the United States was "outmaneuvered." Similarly, the National Association of Wheat Growers asked Secretary Butz to help southern wheat farmers because they sold "following harvest without knowledge of the full impact of the Russian sale." The information problem led the USDA to require exporters to inform the department of large sales and triggered the high-level negotiations that yielded the "U.S.-U.S.S.R. Grain Agreement." The agreement attempted to eliminate the information problem and price fluctuations by requiring that the Soviets commit to a long-term arrangement for buying a specific amount of grain, premised on the Food Deputies Group of the Economic Policy Board's belief that the Soviets were responsible for 80 percent of the price fluctuation in the world grain trade from 1960 to 1975. In 1975, when Soviet purchases were confirmed in advance, Senator Henry Jackson (D, WA) argued that prior notification would "result in earning for American farmers millions of dollars they would otherwise have lost had the Russian grain shopping spree been kept quiet." The agriculture commissioner of North Dakota also noted the information advantage: "The farmer is an avid reader and listener of tips of market or market-related information, but access to adequate information at his farm base is no match for the resources of a grain company with offices and sophisticated communications systems tied to over 3,000 employees going from a field office in our producers area, to marketing, processing, and distribution office around the world."[33]

The Russian grain sales highlighted the importance of the international grain companies, but it also, in retrospect, highlighted the exceptional nature of the circumstances of the early 1970s. Most of the time the surpluses were anything but a help to the grain trade—Cargill almost lost it all twice, as groups like the National Farmers Organization would find out when they attempted to export. Despite the talk of a "grain cartel" and "agripower," the historic nature of the international grain trade was too powerful to overcome. The Nixon administration's attempt to reconstitute the agricultural order—rolling back the domestic farm program, ignoring the IWA, reducing P.L. 480 sales to the Third World, and relying much more heavily on exports—would fail with the return of greater commodity supplies in the latter part of the 1970s and in the 1980s. The critics misunderstood the near-constant disequilibrium in international grain markets that made exporting unprofitable absent state subsidies, the successive failures of International Wheat Agreements to function owing to cheating and noncompliance (problems that would similarly bedevil any collusive agreement between grain companies), and the persistence of state-organized dumping programs like P.L. 480. Most fundamental to the problem was the inability of the United States to persuade any other country to reduce output as it was, leaving the entire burden of propping up global prices on the United States. But even if the United States could have reduced world output and boosted prices, nearly any country in the world could counter with greater production of its own. When President Carter imposed a grain embargo on the Soviet Union after it invaded Afghanistan, for example, other sources quickly filled the void—during the American embargo Australian exports to the Soviet Union grew elevenfold. When world prices collapsed in the 1980s, some farm groups actually advocated the formation of a "wheat exporters' cartel" to raise farm prices. The idea failed to generate much interest for the same reasons that previous collusion was impossible—intense trade rivalries among nations and trading blocs, the persistence of state dumping programs, the ease of stepped-up production by non-cartel members, and the intense pressure from domestic consumer groups to keep prices as low as possible.[34]

ENTERING THE EXPORT MARKET

One of the outstanding features of the American grain trade is private organization, in contrast to the state-run exporting boards in Canada and Australia and the similar state-run institutions in the Soviet Union, Poland, Iran, and Japan. When Congress debated the creation of an exclusive government grain-trading agency after the criticism of the 1972 sales, traders protested, but so did groups like the NFO because they wanted to enter the export market, another indication of competition in the grain trade. They were joined by other farm groups when it seemed as if more money could be made in the grain trade and entry would be possible. Instead of "serving as feeders to the big commercial grain exporters," as Congressman Larry

Pressler (R, SD) described it, cooperatives like Farm-Mar-Co., Union Equity, Agri-Industries, the Indiana Farm Bureau, Farmers Export Company, and the Farmers' Union Grain Terminal Association entered the export business; farmers in North Dakota formed the North Dakota Wheat Pool, allowing them to skip the middlemen and sell directly to foreign importers. From 1968 to 1980 farmer-owned cooperatives increased their share of port elevator capacity from 10 to 21 percent, adding six elevators in the Gulf of Mexico ports alone, and thereby increasing their ability to export.[35]

In the late 1950s the cotton-oriented firm Cook Industries entered the soybean market. In the first sale to the Japanese in 1961 the new entrant lost $14,000, demonstrating the difficulties of trading agricultural commodities. Soon the company purchased an elevator on the Mississippi River to collect commodities for export and was actively sought out by buyers wanting to put more competitive pressure on firms like Cargill and Continental. Cook lined up a large sale to the Soviets in 1974, but the Ford Administration suspended it, fearing the low level of grain stocks and upward pressure on food prices. With 2.2 million bushels of grain ready for export when the sale was suspended and prices crashed, Cook anticipated losing $25 million. The contracts were later reinstated after Soviet-U.S. talks, saving Cook. In 1977 Cook anticipated a bumper crop of soybeans and relatively low prices and sold to buyers accordingly without actually possessing the beans—selling "short." A cold, dry, and long winter prevented the normal use of the Mississippi River barge system and thus increased transportation prices, other speculators guessed prices would rise and went "long," and in the fall the beans burned up during a late-season drought. For fiscal year 1976–77 Cook lost about $90 million, sold off its elevators and other holdings, and sold out to the Japanese firm Mitsui.[36]

In the early 1960s the NFO offered its members an opportunity to consign grain for storage and export into global markets. The NFO loaded its first barge for export in 1970 and soon expanded its number of barge points to thirty-six; President Oren Lee Staley asserted that the NFO has "the operating capability to compete with the two largest grain companies." The 1972 sales to the Soviet Union that the federal government coordinated included 13 million bushels of barley that the NFO had blocked for sale as part of its collective marketing program. In under two weeks in late 1973 the NFO loaded fifteen barges for export and contracted for the sale of forty thousand tons of corn to overseas buyers. In 1976 the NFO exported over 2 million bushels of grain by May, including 765,000 bushels

TABLE 8. NUMBER OF FIRMS EXPORTING

	Wheat	Corn	Beans
1974–75	41	56	39
1979–80	54	77	45

Source: "Market Structure and Pricing Efficiency of U.S. Grain Export System," GAO/CED-82-61, June 15, 1982, 18.

of corn and 700,000 bushels of soybeans marketed in the Netherlands; 150,000 bushels of spring wheat to Belgium; 300,000 bushels of Montana wheat to Japan; and 145,000 bushels of Kansas and Nebraska wheat to Europe in general.[37]

When the NFO began its foray into exporting, according to one of its members, "grain market officials . . . scoffed that exporting was no field for amateurs in the grain marketing business." The scoffing underscored the difficulties of the grain trade that the NFO soon discovered—during one period of the price rises in the early 1970s, the NFO lost $225,000 on soybeans, $250,000 on triticale, and $300,000 on milo. Cook Industries, which also suffered huge losses in the 1970s, sued the NFO for $337,000 in soybean contracts the organization failed to fill. The president of the North Dakota NFO noted the difficulty involved in the exporting process owing to loading and shipping costs. "We have a lot to learn about this ball game," he added. *Business Week* reported in 1984 that Farmers Export Inc., which exported for twelve regional farmer cooperatives, "plunged into the red in 1981." Farmland's Far-Mar-Co subsidiary, according to the president, "never made what you would call a fair amount of money, compared with what we put in." While offering support, the head of the USDA Farmer Cooperative Service warned cooperative officials in a Des Moines speech in 1975 that "exporting is a high risk business." The farmers involved in the cooperative sales and the executives at Cook surely agreed with the NFO president: "It gets a little tough to try to figure out what the dickens to do with the market." Despite difficulties, the NFO continued to participate in export markets and continued to search for international customers in Brazil, China, and Japan, underscoring the ability of farm groups to enter export markets, however risky.[38]

While the experiences of Cook, the NFO, and others demonstrate that the grain trade could be entered, it also underscores the advantages of more established firms. Cargill, for example, was the first to build river terminals, which lowered their shipping costs. They also took advantage of railroad hopper cars, which were less expensive—the cost of a Rock Island

Line fifty-four-car unit train from Des Moines to Houston was 61.5 cents per hundredweight, compared to 74 cents per hundredweight in single boxcars. They also lightened their tax burden by expanding overseas. Since the United States was the only country to tax the foreign earnings of domestic firms, putting U.S.-based firms at a competitive disadvantage, Cargill established Tradax International in Panama, which was not required to pay taxes on earnings outside of Panama. Even with such competitive edges, Cargill lost money in 1968–69, as did other firms in the postwar period when surpluses made it "a time when traders had to fight for a quarter of a cent a bushel"; it even lost money during the Russian grain sales, as did the NFO and others. Even the manager of the GTA admitted that the grain companies "moved a lot of grain, much of it at a loss. We didn't make any money on the Russian wheat deal, and nobody else did." A Cargill vice-president commented that "it's a dog-eat-dog competitive business if there ever was one." The competitive nature of the grain trade, as with the processing sector, casts doubt on the first half of the monopoly problem and highlights the importance of the second half: farmers effectively organizing their marketing.[39]

5

THE NFO AND FARM BARGAINING

In these modern times it is essential that the agricultural sector of the economy balance out with other segments of society. By putting a price tag on farm products fair to the agriculturalist through collective bargaining it would end the vicious cycle of booms and depressions. A lot of unnecessary misery would be done away with also.
ANDREW W. RINIKER JR., FARMER, MASONVILLE, IOWA

I joined the National Farmers Organization as soon as it came to North Dakota. I studied their motives from the start and I believe it is the greatest move ever made in the history of agriculture. It would give us the right amount of free enterprise, using food which is the most powerful weapon in the world, as our power if necessary, to bring the large processors and manufacturers of our products in line with us in the sharing of our nation's wealth. JAMES NILSON, FARMER, SHEYENNE, NORTH DAKOTA

When Dwight Eisenhower occupied the Oval Office in 1953, *Time* magazine declared his two most pressing problems to be the bloodletting in Korea and tumbling farm prices. Eisenhower decided to "go to Korea" himself, but he appointed Ezra Taft Benson secretary of agriculture to handle the "farm problem." The new secretary attempted to replace the fixed farm parity prices of the New Deal and Fair Deal era with "flexible" price supports administered by the USDA, believing it would reduce the incentives to produce and ultimately increase farm prices. But by the fall of 1955 hog prices dropped to a nine-year low; cattle worth $30 per hundredweight in 1951 were worth $11 in 1956; and farmers' net cash income dropped 25 percent from 1951 to 1955. On an average Iowa farm net income dropped from $10,200 to $7,000 from 1953 to 1955. Partly as a result, 3 million

American farmers left their farms during the 1950s. By the time of the 1960 presidential campaign Senator John Kennedy still declared the plight of farmers "our number one domestic problem" and argued that farmers "should not be . . . forced to sell for whatever price [they] can get."[1]

The National Farmers Organization (NFO) originated in Iowa in response to the falling farm prices. After abandoning the hope of a legislative solution to the farm problem, the NFO promoted greater farmer organization and collective bargaining between farmers and food processors so, as Kennedy said, farmers would not have to take "whatever price [they] can get." The NFO achieved a degree of success, stimulated more active farmer marketing and attention to bargaining power, energized other farm groups' efforts to build farmer bargaining power, and stimulated efforts to pass legislation creating a more favorable economic relationship between farmers and processors. The deep divisions among farmers, farm groups, and within the NFO, the hostility generated by NFO tactics, the problems inherent in collective action, farmer independence, and other obstacles to effective bargaining hindered the NFO movement. But despite these problems, the history of the NFO indicates that farmers could organize and receive better prices at times and were not always the hapless victims of the corporate "monopoly problem." NFO achievements also indicate that the organization was not a fleeting movement doomed to failure, contrary to interpretations that hold that the NFO "had little impact on agricultural prices" and presented a "pale replica of the protest organizations of the 1890s or 1930s."[2]

Unfortunately, in postwar American historiography the story of the NFO is overshadowed by Birmingham, Saigon, and Haight-Ashbury. The neglect of this large-scale social and political movement reflects historians' minimal attention to the small towns and rural areas of the postwar American heartland, places that only passively experienced the civil rights movement, war protests, or the counterculture, in which many current historians participated and therefore privilege, and which consequently dominate the postwar narrative. It should be remembered that the tensions in the cities and on the campuses that exploded into rioting and violence had a parallel in postwar rural America stemming from the farm problem and the NFO's efforts to resolve it. On top of being accused of tolerating "violence in the streets" during the long, hot summer of 1964, LBJ was also accused of tolerating "violence on the farms."[3]

The NFO started during a conversation between a farmer, Wayne Jackson, and a feed salesman, Jay Loghry, in Corning, Iowa, in September 1955.

The conversation led to a series of sale-barn meetings of Adams County, Iowa, farmers who decided to form a new farm organization to do something about farm prices. In October eighty farmers from sixteen Iowa and eight Missouri counties met in the Legion Hall in Corning and voted to ratify the articles of incorporation for the NFO. Organizers for the new organization fanned out around the Midwest, asking farmers to pay the one-dollar fee to join, and soon claimed 5,000 members in Iowa, Missouri, South Dakota, and Minnesota. By the time of the first formal meeting in December, the NFO boasted nearly 56,000 dues-paid members, and by the spring of 1956 it claimed over 180,000 members. Daniel Webster Turner, a Republican governor of Iowa in the 1930s, served as a key organizer, speaking in sale-barns around Iowa, Missouri, Kansas, Nebraska, Illinois, Wisconsin, Minnesota, and other states throughout 1955, 1956, and 1957. Turner exemplified the hatred of the Eisenhower-Benson farm policies—he chaired Eisenhower's campaign in Iowa in 1952 but refused even to vote for him in 1956.[4]

Petitioning the federal government for prices at 100 percent of parity became the immediate priority of the NFO. This was also the legislative agenda of the National Farmers Union (NFU), but the NFU was not an effective outlet for early NFO members; the Iowa Farmers' Union broke off from the national organization in the postwar years and ultimately formed the radical U.S. Farmers Association, and the Farmers' Union never organized in Missouri owing in part to the existence of the Missouri Farmers Association. In January 1956 the NFO used its dollars to send its president, vice-president, and two directors to Washington to testify before Congress about the economic problems of the farmer. Dan Turner wrote in the first issue of the NFO *Reporter* that "our purpose and our duty is to bring the power of our organization to bear on Washington to take immediate steps to assure the farmers their share of the national income."[5]

Despite the early success of NFO membership drives, farmer interest waned and the NFO suffered through its "dark days," total membership falling to forty thousand by the time of the 1957 convention. John Lane, head of the Iowa state NFO board, told Farmers' Union officials that the NFO was at the "end of its rope"; an NFU official reported that the NFO was "gasping its final intake of oxygen" and had issued its last newspaper. The money from earlier membership drives drying up, NFO president Oren Lee Staley admitted that the NFO was struggling with the "problem of making the transition from a protest group to an organization." In October 1957 Lane traveled to Detroit to seek funds from Walter Reuther, president of the United Auto Workers union, and advice on the future direction of the NFO. With the return of an Iowa NFU chapter in 1957, NFO president Staley and NFU president James Patton discussed merging the two organizations, as did Lane and Reuther, and rumors of the NFO disbanding spread throughout the grain belt.[6]

In the midst of this uncertainty the NFO adopted its collective bargaining program. At the fall 1957 NFO convention, UAW officer Pat Greathouse, the leader of the farm implement workers, addressed the crowd; the next day NFO members discussed the idea of collective bargaining and voted to create a study committee. Pushed by Staley and his supporters, the 1958 convention adopted the goal of collective bargaining for farmers. Although traceable to these years, only fragmentary evidence exists of the specific steps taken leading to the decision to advocate collective bargaining and downplay government economic aid to farmers. At one of the early meetings of the yet-unnamed organization in September 1955, one of the main organizers mentioned that organized labor might be a good model to follow and discussed a "union" of farmers. At subsequent meetings in Bedford and Clarinda, Iowa, two representatives of the CIO United Packing House Workers of America who worked the Omaha–Sioux Falls and Des Moines area attended, mixing in talk of strikes with the original plan of petitioning the federal government; an agreement that union packers would refuse to slaughter any livestock not sold by farmers showing badges proving NFO membership was discussed at other meetings; and some farmers talked of affiliating with the AFL-CIO. NFO literature argued that relations could be friendly because organized labor does "not want the two million so-called small, inefficient farmers added to the ranks of labor. They recognize the scores of thousands of unemployed workers in the farm implement industries are out of work because of the loss of farm buying power." One organizer for the Machinist union promoted the labor-NFO connection and asked Governor Turner to supervise labor-NFO efforts to talk to farmers and "wake them up." When membership plunged and the odds of the organization surviving narrowed, the NFO received three thousand dollars in financial aid from Reuther and the UAW to help fund the NFO *Reporter*, and Greathouse was soon telling the 1957 NFO convention to forget any chance of help from the federal government. About this time, as NFO literature later described it, "it became immediately obvious to the farmers in the delegation [to Washington] that there would be no change in the gradual but relentless downward pressure on prices by government action (or lack of it)." NFO members began to talk about bargaining under the auspices of the Capper-Volstead cooperative law, what came to be known as the "Wagner Act for farming," and about "decent minimum wages" for farmers.[7]

The NFO's lack of faith in farm cooperatives and doubts about the future of the federal farm program partly justified the collective bargaining approach. The cooperatives, according to NFO vice-president Erhard Pfingston, left a "complete vacuum" in the marketing system precluding a "full, fair price" for farmers' commodities. Benson's and others' political attacks on the federal farm program would ultimately prove successful, the NFO felt, especially given the dwindling political clout of farmers and the mounting clout of consumers. A 1962 proposal by the corporate-sponsored Committee on Economic Development (CED) to phase out the farm program completely in five years in order "to induce excess resources [in other words, compel farmers to leave the land] to move rapidly out of agriculture" added to the NFO's doubts. The appointment of Earl Butz and the formation of the USDA's "young executives" committee in 1971, who both advocated proposals similar to the CED, heightened the cynicism about help from the federal government. A sign on an NFO truck captured the attitude: "To Hell with D.C., Price It Yourself, NFO."[8]

In 1958 the collective bargaining approach became the "main thrust of [the] NFO." Staley wrote President Eisenhower and argued that the "only answer for stability in agriculture is justice at the market place to be established through collective bargaining." NFO vice-president Bob Casper told the Senate that the "lack of bargaining power" was agriculture's biggest problem. Older marketing organizations were nonworkable, as the NFO saw it, because they operated "within and through the existing marketing structure (our outdated marketing system)," which was perverted and manipulated by the coming of large food corporations in the early part of the century. Staley argued that "the law of supply and demand as a basis for getting a fair return for the farmer is outdated. The law of supply and demand could only reflect a fair price to the farmer if the buyer and seller met with equal strength. That situation no longer exists." The NFO thought collective bargaining would fundamentally alter the marketing structure, but as long as farmers were disorganized, "cut each other's throats," and marketed like the "most inefficient slobs on Earth" to the concentrated processing sector, they would fail to improve their lot. Staley asked: "Why shouldn't farmers organize for better prices at the market place? Labor is organized. Lawyers are organized. School teachers are organized. Black people found that through organization they could achieve social justice.

There's nobody below the farmer anymore. He's the low man on the totem pole." The NFO needed to organize farmers in a more businesslike manner, as Vice-President Erhard Pfingston put it, not as a "panhandler begging for a price."⁹

The NFO's diagnosis received support from some economists. In 1952 John Kenneth Galbraith published his book on countervailing economic power, drawing on an economic and political ideology prominent during the Great Depression. As Galbraith saw it, "competition . . . as the autonomous regulator of economic activity . . . [had] been superseded"; concentrated economic power in certain economic sectors required countervailing power in adjacent sectors to serve as a counter. Galbraith cited the model followed by the NFO, organized labor, where countervailing power was "most fully developed." Galbraith devoted an entire chapter to agriculture, where he argued that the market power of farmers was "intrinsically nil" and applauded their efforts to build countervailing power through cooperatives and federal legislation, not specifically mentioning the idea of formal collective bargaining for farmers. In the 1960s, following Galbraith, President Staley would talk about "counteractant power." R. E. Schneidau, an economist at Purdue University, was more specific. He told an audience at the Mid-American Red Meat Marketing Conference that when a "competitive sector butts heads with a less competitive sector—the competitive sector is the loser in the long haul." Because "market prices become a function of relative bargaining power among adversaries," the answer was to build the farmers' bargaining position while using antitrust laws to "remove some of the pricing power of the processors and retailers."¹⁰

Others were not as confident about the economics underpinning farmer bargaining. Peter Helmberger and Sidney Hoos noted that much of the logic supporting the view that bargaining would boost farm prices stemmed from the assumption of a bilateral monopoly relationship between farmers and processors. Robert Clodius noted that bilateral monopoly assumed limited output and no close substitutes existed, an unlikely assumption for such core grain belt commodities as livestock, wheat, beans, and corn. The game theory used to "solve" the bilateral stand-off led to many different conclusions and neglected to weigh ignorance, uncertainty, the impact of negotiations, and other factors. Despite the reservations, Helmberger and Hoos still asserted that in farm markets with elements of monopsony, farm bargaining may increase prices and reduce processor collusion "by altering or threatening to alter the distribution of raw product supplies among buyers according to their willingness to make concessions."[11]

The NFO's plan involved Marketing Area Bargaining Committees, elected by NFO members, presenting offers to meatpackers and promising a steady flow of livestock, all in exchange for contracts for better prices; an agreement required a two-thirds vote of approval from individual County Bargaining Committees. Because packers faced few problems acquiring livestock in the first place, they had little interest in bargaining with the NFO. The result was the beginning of NFO "holding actions," or efforts to keep farm commodities off the market until the establishment of a farmer-approved price. In 1959, a year after the decision to promote collective bargaining, NFO members around St. Joseph, Missouri, held their livestock for an entire week. The holding action failed when the packers in St. Joseph trucked in livestock from outside the organized area. In a larger, more effective hold during the second week of April 1961, the NFO significantly reduced the amount of livestock marketed over the 1955–60 average for that week: Omaha cattle 27 percent, hogs 32 percent; Kansas City cattle 26 percent, hogs 27 percent; St. Joseph cattle 28 percent, hogs 56 percent. Instead of livestock prices rising during the hold, they remained the same; instead of declining after the hold, they increased. Based on these "tests," the NFO concluded that the "so-called free market is a myth—that livestock market prices are rigged on a fantastic scale."[12]

The NFO organized its "most effective" holding action to date in 1962. At the convention in December 1961, the NFO decided against more test actions and called a general holding action for the entire grain belt, a decision ratified in a series of local NFO meetings in early 1962. In August twenty thousand farmers met in Des Moines in an NFO "Meeting for Action" immediately before the hold. On Saturday, September 1, the holding action started; with Labor Day on Monday, three days would pass before the marketing of any livestock. The week before, anticipating the coming holding action, some packers contracted with farmers unaware of the impending action for delivery on the day after Labor Day. Despite this planning, and even though more livestock was marketed on the Friday before the holding action started than during any day in the previous fourteen years, prices surged. As receipts at local and terminal markets dropped sharply, packers began to lay off workers. In the second week of the effort many non-NFO members who had cooperated "began deserting the movement in droves"; on October 2 the NFO called off the holding action. The 1962 effort "totally depleted the finances of the organization and left many local county leaders and members understandably discouraged," delaying further actions until 1964. The only academic article written about the holding actions indicates that the NFO was able to sign three contracts with packers during the 1962 holding action, however. But during the 1964 holding action, which was also held in August for forty-two days and covered twenty-three states, the NFO claimed to get the first "verbal agreements" from some meat processors to buy from NFO members. The same document indicates that during the holding action in 1968 the NFO was also able to arrange the first written supply contracts based on pricing formulas pegged to market price levels. Another document indicates that "1963 has really been the year in which the economic barriers for farmers have been broken by the NFO" with the signing of "master contracts" with processors. The American Meat Institute told its packer members in October 1964 that no packers had signed contracts with the NFO and said it "was unfortunate that some farmers are still being misled by the NFO's false statements."[13]

The NFO believed that the successes of their holding actions were underplayed by the USDA, the press, and the processors in order to undermine their efforts. While NFO members reported packers' pens empty and receipts down to a "trickle" during holding actions, the Federal Market News Service reported "normal" receipts, and processors reported little affect from the holding actions. The NFO denied that the holding action served to increase supply and further depress prices by keeping livestock ready for slaughter off the market. Instead, they argued that the additional weight added by livestock held off the market was offset by the hogs that were slaughtered twenty to fifty pounds light and the cattle slaughtered three hundred to five hundred pounds light. They added that the farmers who held livestock off the market didn't push for maximum weight gain and kept rations to a minimum.[14]

How much effect the holding actions had on prices, and when and how many contracts were signed and for what price, is hard to know. Fragments of evidence from the NFO collection, however, do seem to indicate NFO successes. The manager of the Agricultural Research Division of Swift and Company wrote to Staley indicating he could not sign a master contract with the NFO that failed to adjust to situations when prices were falling. He did express interest, however, in continuing to buy the livestock of NFO members on a competitive basis. In 1972, according to the NFO, three "major Midwest packers" agreed to sign one-month contracts paying "NFO members not less than a fixed cost-plus-a-profit figure for an assured supply of hogs." When the number of these contracts grew to twelve, a "top official at one of the biggest packing companies" said that "this is the biggest thing NFO has ever successfully undertaken." The NFO papers also include "letters of intent" to bargain from Spencer Foods, MBPXL (Missouri Beef), Certified Grocers of Illinois, Borden, and Beatrice. In the late 1970s the NFO told Congress that the prices NFO cattlemen in Wisconsin received from packers were markedly higher than the Wisconsin Statistical Reporting Service State Average and that they had contracts with Armour, Wilson, Morrell, Swift, MBPXL, Dubuque, Sioux Pac of Iowa, and Packerland of Green Bay. In the mid-1960s the Kiplinger financial letter reported that meat packers "privately" gave credit to the NFO for boosting hog prices four dollars and cattle prices two dollars. Labor also offered credit. When the Hormel plant in Austin, Minnesota, cut back production because of a "contract problem," the president of the local union disagreed, saying it was because "Hormel is not getting its share of the available supply of livestock, in part because it has not established good enough relations with the NFO." Instead of Hormel receiving the one thousand hogs sold at the local NFO collection point every week, they went to Swift in St. Paul and Morrell's in Sioux Falls and Ottumwa; NFO cattle went to the Armour plant in St. Paul.[15]

The Rath Packing Company, ranked eighth in national gross sales with plants in Waterloo and Columbus Junction, Iowa, offers more evidence of the effectiveness of the holding actions. During the 1962 action, the reduction in supplies forced Rath to lay off two hundred workers at the Waterloo plant. After the 1964 holding action, internal Rath memorandums conceded that an "NFO withholding action can be very effective over the short term" and noted a sharp advance in prices during the 1964 action. Anticipating a holding action in late 1967 and early 1968, the company slowed its sales to retailers, canceled advance bookings and previous arrangements based on the "yellow sheet" (Federal Market News Service) or the Chicago market, and prevented salesmen from quoting prices to retailers, fearing that prices would rise dramatically if a holding action came. They also decided to "approach NFO after a day or two if their action is aimed at Rath to see if something can be worked out" and to "listen if they do approach Rath with discussions." In the spring of 1968 Rath reached an agreement with the NFO that they would pay a "local country price" for hogs originating from the NFO's country collection points plus fifteen cents, five cents for arranging the delivery and ten cents for the collection point operator; Rath wanted twelve hundred hogs daily.[16]

At some point in the 1960s the NFO emphasis switched from large-scale holding actions to block marketing, or the pooling of NFO members' commodities for sale to processors at a negotiated price. As one piece of NFO literature described it, blocking was a way of incrementally building market power and bargaining with processors until enough commodities were blocked that farmers could set their own price. With the block marketing program the NFO could "lift" commodities out of regional markets when prices were higher in others. For example, the livestock market in southeastern South Dakota was dominated by the Morrells packing plant in Sioux Falls. Instead of selling to Morrells, NFO members in the area blocked together nearly 60,000 hogs and sold them in Omaha for a higher price. Other scattered citations from the NFO papers indicate a great deal of success at blocking. Agreements on hogs in 1972 were partly supplied by a 100,000-hog block the NFO built along the Missouri River. A series of hog collection points for "Operation Little Porker" started in Strington, Iowa; Elizabethtown, Kentucky; and Madison and Redfield, South Dakota, and grew into a system collecting nearly 2,000 hogs every few days for sale to packers. The head of the NFO meat department claimed that on some days the NFO collection points marketed more than all the terminal markets combined. One day in 1973 the collection point in Ghent, Minnesota, marketed over 1,000 hogs and received $4 over the market price for them; for the year, the point marketed over $1 million in hogs. A high-ranking NFU official admitted that he was "quite impressed by the efficiencies which NFO has introduced in marketing of livestock . . . [which] have resulted in some increased income for NFO members." In 1973 farmers in northeastern South Dakota blocked 100,000 bushels of corn for sale in Duluth, Minnesota. Since the local South Dakota elevators were buying at $1.36 a bushel and buyers were paying $1.87 in Duluth, the farmers ended up receiving about $1.57 for their corn after discounting the transportation costs. When the farmers blocked another 136,000 bushels, the local elevators boosted their prices to $1.85 a bushel. In the Madison, Wisconsin, area farmers blocked corn and sold it on the Illinois River Market for up to $2.10 a bushel, compared to the local elevators' price of $1.57. Farmers around Cherokee, Iowa, blocked another 100,000 bushels of corn and transported it to St. Joseph and Kansas City for loading onto Mississippi River barges for $2.10 bushel, compared to the local price of $1.12. Staley contended in 1977 that the NFO had "thousands of contracts" in effect with packers, grain dealers, and milk processors and that "eight of the nation's 15 largest meat processors are accepting production from NFO members as such." An NFO brochure indicates that million-head blocks of cattle were organized, and another includes on its list of outlets Hygrade, Armour, Dubuque, Spencer, Wilson, and Hormel, along with others. In 1973 NFO signed a contract to supply the MBPXL (acquired by Cargill later in the 1970s) plant in Rockport, Missouri, with 70,000 cattle per year.[17]

In addition to these fragments of evidence, a version of the building of the NFO marketing structure can be gleaned from the 1969 book *Angry Testament*, written by the NFO partisan Charles Walters. According to this account, after the 1964 holding action the NFO "erected the most fantastic marketing structure ever attempted by a farm organization." It started with the cooperation of an official of the Wabash railroad who invited NFO members to begin collecting hogs in Griggsville, Illinois, for NFO-sponsored shipment to "where the market was and the hogs weren't." From this beginning grew the wider system of NFO collection points and NFO marketing practices that, according to Walters, "were delivering a buck or two a hundredweight above prices available from packer agents or terminals." Testifying before Congress in the late 1970s, the agricultural economist Harold Briemeyer reported that the NFO "repeatedly makes sales at $2 over the sheet."[18]

In addition to marketing strategies, the NFO altered the political economy of agriculture in other ways. At times, the NFO completely bypassed the traditional marketing channels and sold "farm fresh" commodities directly to consumers through "NFO farmer-to-consumer" sales. In 1975 the NFO estimated the level of daily sales to be twenty thousand pounds of cheese, seventy-two thousand pounds of ground beef, and forty thousand pounds of potatoes. In the late 1960s the NFO also lent money to the Glenwood Packing Company to save it from bankruptcy; when the bankruptcy left many farmers unpaid for marketed meat, the NFO started offering members insurance against such possibilities. The NFO wanted to keep the company afloat because it had bargained with them, but the effort failed. The NFO established its own grain terminals and leased over six hundred hopper cars as part of its nationwide grain-marketing system; one NFO publication indicated that from January to May 1973 the NFO signed 949 different grain contracts. The NFO even organized farmers to sell millions of bushels of grain at precisely 2:35 P.M. on the same day all over the grain belt, preventing the Board of Trade from adjusting to the increased supply. The NFO emphasized marketing education; in the 1970s the NFO Hog Division guided farmers through a forward-contracting program for hogs as another alternative to the normal blocking and bargaining program. The

NFO also broadcast its weekly radio show "Here's Info" to promote its efforts, broadcasting on forty-five stations in the corn belt in 1966 and expanding to over twelve hundred stations by 1978.[19]

Although the NFO kept total membership a closely guarded secret, a lawsuit in the 1970s provided information that allowed observers to estimate membership at between 65,000 and 175,000. It also disclosed that NFO income grew from about $600,000 in 1960 to $7 million in 1971, much of which was used for salaries, travel, and office expenses for NFO staff. The income stemmed from dues and from checkoffs on NFO sales. In court documents the NFO "estimated" it marketed $562 million worth of commodities from October 1969 to September 1970. In 1973 Staley maintained that the organization marketed over $1 billion in commodities. Another newsletter obliquely points out that NFO conducted a dozen times as much business as the Farm Bureau, which would equate to about $3.3 billion in farm marketing a year.[20]

The formation of the NFO and the decision to advocate collective bargaining also changed the center of gravity in farm politics, stimulating similar interests among more established farm groups. The American Farm Bureau Federation (AFBF), fearing the successes the NFO might have and the resulting effect on AFBF membership, decided in the late 1950s to seek a greater level of bargaining strength for farmers by forming the American Agricultural Marketing Association (AAMA). Smaller efforts also developed, such as the Cattle Marketing Association for Farm Bureau members in South Dakota, which was established to sell collectively through the Producers Commission Association in Sioux City. In addition to fearing the NFO, the AFBF became convinced that more than the Capper-Volstead, Packers and Stockyards, and FTC Acts were needed for them to effectively bargain. This view partially stemmed from events surrounding tomato marketing in Ohio in the 1950s. In 1950 a group of Ohio tomato farmers formed Cannery Growers Inc., hoping to sign production contracts with processors at higher-than-market prices. In 1952 the FTC ended up charging such tomato processors as H. J. Heinz, Hunt Foods, Campbell Soup, and Stokely–Van Camp with boycotting the organized growers. After several twists and turns within the FTC, a U.S. Court of Appeals dismissed the FTC's cease-and-desist order in 1959. When harassment of Cannery Growers' members continued, the FTC admitted that it didn't have the authority to stop the processors' "coercive tactics," but only to prevent combinations and conspiracies among processors. Because of these problems, the tomato growers who were affiliated with the AFBF sought state legislation to protect themselves from processor abuses. In 1965 the legislature passed a law establishing a complaint procedure for farmers concerned about processor activity and forbidding processor boycotts.[21]

The status of broiler growers in the South also justified AFBF's interest in bargaining legislation. The Northwest Poultry Growers Association was formed in response to concerns about the situation of farmers contracted to feed and raise broilers. Companies such as Ralston Purina and Tyson refused to buy from the members of the association, however, and the number of grower members dropped from 313 in 1962 to 23 in 1963. After investigations by the Packers and Stockyards Administration determined that processors were intimidating members of the association, the government ordered all processors to cease and desist from boycotting, intimidating, or blacklisting growers. Randall Torgerson concluded that the AFBF's view that previous laws were enough to foster effective farm bargaining efforts "were demolished by experiences with the FTC in the Ohio tomato case" and by the "unresolved experience of broiler growers in Arkansas."[22]

To improve the farmers' bargaining ability the AFBF sought legislative alliances with the National Council of Farmer Cooperatives (NCFC) and the National Milk Producers Federation (NMPF). These groups also sought to build farmers' market power and to limit the effectiveness of the NFO, which threatened their long-standing marketing institutions and which might lure away members. Unlike the AFBF, the NCFC and the NMPF were reluctant to tamper with the provisions of the Capper-Volstead Act, fearing that attempts to amend it might lead to other unintended consequences. Specifically, they feared the power of the National Tax Equality Association, which sought to repeal cooperatives' tax-exempt status throughout the postwar period. Nevertheless, the coalition decided to find a congressional sponsor for a bill to amend the Capper-Volstead Act to afford the AAMA the same status as cooperatives under the act, to legalize affiliation with the AFBF, to allow state farm bureaus to combine and coordinate commodities for sale, to exchange information on sales, to expressly include the term "bargaining" in the act, and to forbid processor manipulation of growers. The AFBF, NCFC, and NMPF, however, also feared that the NFO would be strengthened by such legislation, a fear that led to counterproductive legislative and political wrangling. The assistant legislative director of the Farm Bureau asked, "Why are we for this kind of bill when it would help NFO?"[23]

When congressional leaders introduced the AFBF-NCFC-NMPF coalition-written bill in May 1964, one of the immediate concerns raised by members of the NCFC involved the NFO. Fearing the bill "could very well provide a real boon for groups like the NFO," some proponents argued that the "associations" protected in the bill should be defined as those who actually assumed ownership of commodities, which the NFO did not. Although disputes continued within and between farm groups about the wisdom of a bargaining bill and the form it should take, the same bill was reintroduced in 1965, this time with the support of the National Farmers' Union (NFU). When Senate hearings on the bill opened in June 1966, the AFBF aligned with the NFU and sounded like the NFO, emphasizing the growing power of processors through merger and diversification and the reprisals against farmers who attempted to organize bargaining associations. Divisions among farm groups, the resistance of processors, and the desire to pass some legislation prompted sponsors to deemphasize "bargaining" in favor of a more neutral "fair-play" bill, subjecting cooperative processors to the same provisions as corporate processors. In the next Congress, differences over the specifics of the legislation among farm groups, the opposition of processors, and the icy response from the USDA, the Department of Justice, and the FTC resulted in legislation considerably less forceful than many of the proposals discussed, legislation that simply prohibited "unfair practices" by both processors and farmer groups. The NCFC argued that the bill had taken on a "strong anti-cooperative and anti-farmer affect and tone," and the NFU saw the bill as a "mere shadow of what it had been." The NFO, which supported the earlier legislation, urged a presidential veto; an NFO supporter complained that the bargaining bills in Congress had been "amended half to death, fed cancer cells that would explode in the veins of the primary producer, so that they could finally deliver to the handlers and disposing end of agribusiness the real market power."[24]

Opponents viewed the bargaining legislation as more than a simple "unfair trade practices bill." They feared the coming of a regulated bargaining regime similar to that used by organized labor. Some chicken processors supported the modified legislation by arguing that if no law passed that would help farmers to organize, then labor unions or the federal government would step in to do it, an event "tantamount to inviting—tomorrow or the day after—a full scale Wagner Act for agriculture." Secretary of Agriculture Orville Freeman conceded these arguments when he supported the bill as a step toward improving the bargaining position of farmers, pointing out that wider legislation based on either the Wagner Act or the

National Industrial Relations Act might be appropriate, and talked of supporting a National Farm Bargaining Board, marketing orders, and producer marketing boards, much like the National Commission on Food Marketing report recommended in 1966. The leadership of the Grange argued that a provision forcing processors to "bargain in good faith" should be included in the bill, alluding to earlier decades when labor unions formed but management still refused to negotiate. Again following earlier NFO ideas, the AFBF also considered legislation similar to labor laws, including provisions allowing injunctions against processor interference. The NCFC also desired a Wagner-type proposal including a closed shop and the compulsory recognition of farm bargainers.[25]

In 1968 Senator Walter Mondale (D, MN) introduced, with nine grain belt Senate cosponsors, the National Agricultural Bargaining Act—unveiled by the NFU president at the National Farm Institute in Des Moines—which invoked many of these ideas in an effort to "create a national collective bargaining system." Specifically, the bill established a National Agricultural Relations Board, which selected a marketing committee for each farm commodity. An election (simple majority) would determine if producers wanted to bargain. If they voted to bargain, a negotiating committee would be selected to bargain for a price with processors, circumventing the open-market and cooperative system of marketing. All producers would receive, and all processors would pay, the price agreed upon during negotiations. Mondale believed that the price demands of the farmer committee would not be excessive owing to "serious risk of competition from substitute foods, increased imports, or, in the absence of supply control, tremendous production increases." Title 2 of the bill amended the Agricultural Agreements Act of 1937, making grain belt commodities eligible for federal marketing orders like those that existed for fruits, vegetables, and milk. This model, often discussed by grain belt farm groups and politicians, provided for a farmer referendum (simple majority) on a marketing order establishing total output and minimum price.[26]

Wagner-type legislation for farmers was a hard sell. The cooperative tradition always touted voluntary membership, and the "antidemocratic" implications of compulsory membership in bargaining associations alarmed some farmers. Although many farmers complained about the monopolistic processors, their "propensity to regard agriculture as an integrated whole, inclusive of processing and handling, [made it] doubtful that the essential feeling of militancy and solidarity exist[ed]." Unlike industrial workers, the interests of farmers were more varied, as the many battles over bargaining legislation and farm bills indicated. A political operative surveying farmer opinion on the plains in 1972 concluded that bargaining legislation was opposed because of the "threat of mandatory controls. Mandatory anything makes [farmers] mad." The Farm Bureau objected to the mandatory participation (if the referendum passed), the producer record-keeping requirements (and the five-hundred-dollar fines for failing to keep records), the provision requiring the national board to choose bargaining committee candidates from a list prepared by the Agricultural Stabilization and Conservation Service Committees, the compulsory arbitration provision, the potential for strict production controls, and to further enraging inflation-conscious consumers, a growing power in Washington. Some cooperative advocates feared that bargaining committees would displace cooperatives or that processing cooperatives might be designated as "processors" in the bargaining process and also resisted the price-fixing and mandatory membership provisions. When Senator Allen J. Ellender (D, LA) wondered whether the farmer bargaining committees would constitute a "superstructure" over cooperatives, proponents could not explain the relationship. The fiercest resistance came from processors: a processing company official in Wisconsin predicted a "police state in agriculture" and the "socialization and Russianization of our economy." And the Washington Post, in a period of greater and greater calls for antitrust action, concluded that the bill was "basically a price fixing bill."[27]

The April hearings on the Mondale bill deepened the divisions in the farm community over such legislation and highlighted the resistance to the level of control needed to make the law function. The bargaining issue received some attention in the spring primary season—in his South Dakota and Nebraska wins, Robert Kennedy listed bargaining legislation as the first thing he would seek for the farmers as president. Events eventually overshadowed the legislation, however—Tet, LBJ's decision not to run for reelection, the assassinations of Kennedy and Martin Luther King, rioting, the run on gold—and beginning in the summer of 1968 the Johnson administration and the USDA played down the bargaining program as an effective approach to the farm problem. During the presidential campaign, agriculture was not a high priority for Richard Nixon: "The national press couldn't care less about what we say on Karl Mundt's pet REA project nor on our repeating our agriculture program. In fact, I think the less we speak nationally on agriculture in the next few weeks—the better." In Crete, Nebraska, only weeks before he was assassinated, Kennedy asked the crowd if the other Republican candidate was any better: "Do you think Nelson

Rockefeller understands the problems of the farmer?" When the Nixon administration came to power in 1969 the chances of a bargaining bill's passing dwindled further. Secretary of Agriculture Earl Butz, while tacitly supporting a bill to improve bargaining power, made clear his opposition to a "union shop" and antitrust immunity for the bargaining associations, and his support for a time limit on "good faith" bargaining, as did the Department of Justice. Although the administration recognized the "enormous political implications" of the legislation and the political debts owed the Farm Bureau, it gave weak support to the effort. A "high administration source" told the *Washington Post* in 1971 that "if you look at our proposed qualifying amendments [to the bargaining bill], you'll see there really isn't much left." By 1974 the USDA's Clayton Yeutter was still saying that procedures for building farmer bargaining power still had "not been settled." Antitrust officials in the Justice Department also feared the tendency of legislative proposals to "confer upon an approved association a monopoly of the handler's business, a monopoly of the area's produce, and a monopoly of the producers' memberships."[28]

Even the NFO, which spearheaded the postwar debate on farmer bargaining, opposed Mondale-type legislation. Staley denounced the idea of a "czar over agriculture" who would determine bargaining representatives and bargaining rules as a "closed shop in agriculture," maintaining that the Capper-Volstead Act and subsequent court cases provided enough protection for farm bargaining. The NFO's Washington lobbyist told a congressional hearing in 1972 that "we certainly do not want a politically-biased board appointed by the president or anyone else set up to license bargaining agencies; we want the bargaining left in the hands of the farmers themselves, acting directly or in groups such as our NFO-group." The NFU, typically the greatest supporter of government action, even acknowledged farmers' shrinking faith in any process coordinated by the federal government, noting the "preference for greater reliance on their own organizations and activities in the private sector, such as the bargaining associations and other producer cooperatives which are now operating or contemplated by various groups of producers." When the issue reemerged in the late 1970s, the NFO explained the enormous effort they made to secure bargaining contracts, all of which would be undermined by a government-directed bargaining process. Since "accredited" groups were entitled to the same contract terms as other groups, any "little dissident group" could get the same terms as the NFO, potentially leading to the termination of the favorable contracts the NFO worked years to negotiate.

The NFO also continued to fear that the new authority granted the USDA under the bargaining laws would lead to its abuse as a "political weapon" and that the new regulations would help the "larger, more powerful participants simply because they have the resources to employ lawyers, accountants, economists and other talented personnel." Grain belt support for a bargaining bill also seemed to dwindle: when a House subcommittee voted to kill the bargaining bill, the kill votes came from such grain belt states as Minnesota, North and South Dakota, and Illinois.[29]

The divisions exposed in the legislative battles revealed many of the organizational problems farmers faced. The chronicler of the battle, Randall Torgerson, concluded that the "falling out of farmers and their organizations, and not the lobbying activity of the processors and handlers, had brought the progress of the [legislation] to a halt." While the efforts of the NFO and AFBF indicate that farmers were able to organize to a certain extent, bargain for better prices, and receive some political support, their position was weakened by the many divisions they faced. For example, the idea of "collective bargaining" and "unionism" was not popular in rural areas. Some farmers believed that their production costs grew steadily after World War II—aggravating the "cost-price squeeze"—because of the high wage demands of such unionized sectors as the tractor industry. Others bridled at the widespread corruption in labor unions disclosed by the "rackets" committee hearings during the 1950s. Governor Turner, who gave enormous support to the NFO in the formative years, believed that "a close hook up with organized labor (Pack house workers) sound[ed] a little pink" and the idea of union packers checking for NFO badges "practically anarchy"; a Wisconsin farmer noted that NFO members "continually get sneers of others connecting us with communism"; an Iowa NFO sign was painted over with a hammer and sickle.[30] Senator Bourke Hickenlooper (R. IA) thought the NFO policies "stimulated by Labor Union organizers," and a farm magazine thought the NFO was part of Reuther's plan to dominate the country. By the mid-1950s the Farmers' Union, which promoted farm-labor cooperation after World War II, thought that any affiliation with organized labor courted disaster, recalling that the NFU had been criticized for years for its alleged connection to John Lewis of the United Mine Workers. Many farmers and other supporters, including Governor Turner, quit the NFO over the labor connection. Labor returned the suspicion: Reuther faced great pressure not to work with farmers, and Ray Mills of the Iowa State Federation of Labor made it clear that "we have no place for farmers." The Iowa chapter of the NFO concluded in the 1950s that "negotiations with organized labor proved that no major help would be received."[31]

The social tensions generated by the NFO and its tactics undermined its efforts more than the connections to organized labor. During holding actions, NFO members wanted non-NFO members to join the organization or at least to hold their commodities off the market, something they made very clear. The trucks of NFO members conveniently suffered mechanical failure in front of loading chutes at buying stations. At the Sioux Falls stockyards, farm wives and children stood in the chutes to prevent the unloading of livestock. Sixty NFO pickups carrying one hog each crowded into the loading ramps at the Rath plant in Waterloo to prevent large sales to the company. Farmers who marketed during holding actions were "blacklisted" by the NFO and many times found their fences cut into little pieces the next morning or had dynamite set off in their farmyards. One NFO member sued the sheriff of Poweshiek County, Iowa, for "falsely arrest[ing] and imprison[ing]" his six-year-old son (he was taken to the principal's office) to question him about fence cutting. In Iowa, the NFO manned over four hundred checkpoints to keep an eye on packing plants and the farmers who tried to market to them. In one incident, a farmer trying to market his livestock saw the NFO spotters at a farm market and turned around to head to another market twenty miles away; when he arrived, the same NFO people were waiting, having taken a short cut to get there first. The farmer then took his livestock home. In 1964 NFO members from three Minnesota counties gathered as the holding action proceeded. When they received word that a farmer was trying to sell, they raced to his farm to blockade his driveway. The trucks left by another exit, but later one of the truckers was pulled from the cab of his truck and attacked. Because the NFO spotters often had radios tuned to the frequency the packer's buyers used, they could often get to the farm before them and persuade the farmer not to sell his livestock. In another incident, a packer buying agent insulted some NFO members, and they threatened to hang him; by the time they had gone to their pickup, fetched their noose, and thrown it over a rafter of the buying station, the buyer had bolted out the door.[32]

The violence escalated to the point where the FBI became involved in tracking the incidents. In 1964 five trucks moving livestock from Fargo to a packing company in St. Cloud encountered NFO pickets; of the two trucks that went through, one had a brickbat thrown through the windshield and the other had its cattle let loose. A truck going to unload in Pipestone had a plastic bag full of crankcase oil thrown on the windshield, blinding the driver. Two trucks hauling hogs from Iowa to Sioux Falls drove over bars with sickles welded to them on the highway; one truck lost seven tires and crashed in the ditch. A truck with a load of lambs hit a similar device in Wisconsin and tipped over. Dozens of rifle shots at trucks and trains, dynamite blasts at truck weighing scales and at Oscar Mayer plants, the poisoning of milk being sold, and other cases of "NFO madness" and "guerrilla warfare" were also reported. In response to these threats, farmers and truckers carried guns with them to deter NFO activities and on one occasion threatened to tar and feather NFO members who were responsible; some farmers set up false NFO meetings and beat up the NFO members who came; another farmer waited in his ditch all night with a shotgun in an attempt to catch fence cutters. In Wisconsin in 1964 a trucker ran over and killed two NFO farmers who were picketing the sale of farm goods. Others protested the NFO tactics by forming groups such as the Clarke County Citizens Protective Association in Iowa and the Citizens of Nodaway County Organization for Free Enterprise in Missouri.[33]

The violence and radicalism of the NFO did serious damage to its credibility and undermined its larger goals. The *Des Moines Register* called the "intimidation, violence, and vandalism" "intolerable"; they concluded that the NFO had to prove it is not a "mob before it can accomplish any results for its members." Instead of concentrating on promoting organization and securing better bargaining legislation, the Johnson administration filed a suit during the 1967 milk holding action barring the NFO from "threatening, intimidating, harassing or committing acts of violence" against processors, truckers, and other farmers. Fifteen farmers who didn't join the NFO in one Minnesota County indicated they didn't because it was "too radical" and because they "didn't believe in tactics," "lies, violence, destruction, and turning neighbor against neighbor, friend against friend," and "violence by radical members," among other reasons. A letter to a local paper from one South Dakota farm couple living near Madison stressed the anxiety the NFO caused among longtime friends, neighbors, and relatives and asked, "Do we really want our ideas and opinions to be those of others so much that we resort to force to try to change their minds?"[34]

The NFO also promoted the destruction of commodities to protest price levels, further damaging its image. Massive milk dumpings took place during the milk holding action in 1967; in Bloomer, Wisconsin, a farmer drove his milk truck through town with the valves open, dumping milk in the streets. When the massive hog shoots started around the Midwest in 1968 the NFO's reputation suffered even more. The chair of the Des Moines County, Iowa, NFO resigned in protest. Father Edward W. O'Rourke, the leader of the Catholic Rural Life Conference who always enthusiastically supported the NFO, also protested the hog killings. In 1974, after 650 calves were stabbed, shot, and dumped into trenches in Curtiss and Wisconsin Rapids, Wisconsin, President Ford called it "shocking and senseless." An Iowa newspaper editorial commented that "the whole solemn, gloomy sequence of events suggested the weightiness of the sacrificial rites of the Old Testament where the Hebrews offered lambs and calves as a means of getting them through grave periods." A Wisconsin man decried the "massive, inhumane slaughter of livestock" and argued that the NFO's "action has a parallel in the rioting in our cities, where to correct one evil another is committed." On his radio program Paul Harvey said the pig killing and milk dumping was "encouraged by Walter Reuther."[35] Senator Dick Clark of Iowa, a liberal senator with a solid record on farm issues, called the NFO a "wild group" and President Staley a "wild guy," and a Wisconsin man depicted Staley as a "deranged" "demagogue" who caused "only ruination and strife among good neighbors," both summarizing the views of many grain belt farmers reluctant to join such a group.[36]

If the labor connection and the violence didn't keep farmers from joining the NFO, their independence might. As one farmer commented about the NFO recruiting efforts, "I told them [NFO] . . . I was a farmer so I could be my own boss. If you belonged to them, they would tell you when you could sell your hogs or cattle." One farm wife who particularly hated the violence lamented that "farmers just can't pull together." Reasons for not joining included fears of internal battles similar to those that divided labor unions, concern about the distance from the local members to the national NFO leadership—which undermined loyalty and participation in farm organizations—and worries about the degree of control NFO had over members, what Senator Hickenlooper called NFO's planned "pattern of dictatorialness over farmers." Farm couples, though they often acknowledged that farmers were "stubborn individualists" in surveys, often reached the same conclusion: "We felt that we wanted to be independent." A less diplomatic recollection: "I'll tell you what the problem is. It's us farmers, our own damn selves. When I was trying to control production through the NFO, I learned. A farmer won't give up an inch of freedom for the good of the group. So let him hang separately."[37]

Incentives to avoid participation in the NFO effort, given the uncertainty of others' actions and doubts about the effectiveness of the overall effort, explains why a "Farmers' Wagner Act" with a "union shop" forcing farmers to join bargaining associations was so often discussed as the only way to eliminate "free riders." Farmers strongly resisted such controls, however. Even when farmers were given the opportunity to vote on government-imposed acreage controls in 1963, knowing that the "free rider" problem would have been substantially reduced, they voted no. As George Stigler has noted, the ride may not have been free, but it was relatively cheap, reducing the probability of participation in the collective action and reducing its size, therefore its effectiveness, which reduced participation even further. As a Burt, Iowa, farmer saw it, the NFO and its program was a "'wild goose' chase," a "pipe dream—doomed to eventual failure and extinction," a common attitude that kept many farmers from joining the effort. A poll taken a year after the farmer's comments indicated that only 38 percent of Iowa farmers would participate in an NFO holding action. Of the ninety farmers who were members of the NFO in Boone County, Missouri, sixty eventually dropped out when they came to believe that the NFO plan would not work. A study completed in the 1980s indicated that less than half of farmers were willing to bargain; only 23 percent would join a bargaining association if membership involved controls on production.[38]

The NFO also suffered from organizational problems and internal divisions, constantly struggling to hold the organization together and keep dues paid. One NFO "minuteman," who was part of a network of farmers who could quickly spread the word about a holding action, sale, or the monitoring of packing houses, quit because his neighbors laughed at him for getting involved in a cause they considered hopeless. Dues collection was also a problem, as one NFO member made clear: "Oh, the excuses we heard! From I didn't understand it was a 3 year [contract] agreement, to I lost my shirt in the holding action, to I lost my ass in the holding action, or I was threatened when I joined, I was drunk when I joined, my wife will leave me if I pay, the bank is selling me out next week, I don't have the money, I will as soon as I can, to a few who said 'get the hell off my farm and don't ever come back.'" Organizing the marketing of the NFO members also proved complicated. The Iowa state NFO chairman: "I used to be able to put 1,000 bombers over Germany easier than I get 100 farmers to market their hogs on signal during a given day." The Coast Trading Company sued the NFO for forty different contracts that the NFO failed to fill. The worst case came when farmers who were supposed to market for the NFO in 1972 refused and left the organization with huge contracts to fill. The NFO instead bought the commodities on the market, which was highly inflated owing to the Russian grain sales, costing the organization $20 million.[39]

Such huge losses underscored the financial problems of the NFO. From the beginning the organization faced difficulties, as the aid from the UAW demonstrated. In the early 1970s the Securities and Exchange Commission charged the NFO with borrowing over $7 million from members while "employing deceptive devices" and "untrue statements" about the true financial status of the organization. During the investigation, the SEC concluded that the NFO had $17 million more in liabilities than assets and revealed that it was counting $47 million in "back dues" as assets. The NFO faced further problems when it attempted to coerce back dues from farmers, spreading the word that the organization won every court case involving dues it ever fought and telling farmers that if they refused to pay they would have to pay double the amount owed. Such tactics resulted in an injunction against using such statements from the Iowa attorney general.[40]

Other legal problems faced by the NFO were much more serious. When the organization started, it needed to insure that it was organized in such a way that it fell within the provisions of the Capper-Volstead Act. It also had to spend large sums of money to insure that its contract with farmers was legally sound. The NFO battled the USDA to be defined as a "service organization" instead of a "market agency," which involved additional regulation and bonding requirements. The antitrust thicket also complicated the NFO's efforts. In 1962 government antitrust officials said they would closely examine any contracts that may be signed between packers and farmers, further chilling packer willingness to bargain. In 1967 the Department of Justice filed antitrust complaints against the NFO for restraining trade between NFO members and nonmembers and curtailing interstate milk marketing during its milk holding action, lodging a "bone in [Staley's] throat" that LBJ had to personally dislodge to prevent the annual NFO meeting from turning into an "anti-administration rally." The same holding action triggered a massive legal war between the NFO and Associated Milk Producers Inc. (AMPI) and the Mid-American Dairymen cooperatives. The co-ops charged NFO with "conspiracy in restraint of trade," a Sherman Act violation, for its efforts to build a large block of milk producers; AMPI documents revealed they believed a large, long, expensive antitrust battle would "'break' NFO's back." The NFO countersuit, resolving questions of market definition, market power, proof of intent, and the agricultural exemption from the antitrust laws, led to a twenty-year-long legal battle. The battle between groups who were supposed to be helping farmers became so frustrating and destructive to farmers' interests that some religious leaders intervened and attempted to broker a settlement.[41]

On top of doubts about the NFO's program and legal and financial problems, the NFO's credibility suffered from internal infighting. Many resisted the labor contribution—which triggered the resignation of the national secretary and treasurer—and soon after, the entire Iowa Board of Directors voted to merge with the NFU. Staley fought the National Board on the collective bargaining idea, which such prominent NFO officials as Vice-President Bob Casper opposed. Staley was opposed in a bitter presidential race in 1962 involving charges of his authoritarianism, and after his victory most of those who failed to support him were purged from the organization. The 1972 convention involved the "most vicious, the most emotional, and the most destructive internal political movement that Staley and his supporters ever encountered." The battle involved the dismissal of six members of the National Board, who were seeking to oust Staley for allegedly breaking NFO rules. Charles Walters, an NFO supporter who had written two books about the organization but subsequently resigned in 1972, sided with the anti-Staley forces. Walters compared those who failed to oppose Staley to Erwin Rommel, since they did "not wish to dispute the Fuehrer." When the six men were expelled, Walters concluded that Staley "survived again, just as he had survived before, by piling men and their reputations higher than corpses at Shiloh's breastworks." Large segments of the NFO began to distrust Staley, particularly members from the newly organized areas of the South and West. One NFO partisan even claimed that attempts on Staley's and Erhard Pfingston's lives were made that the FBI investigated. Whatever the merits of the charges leveled back and forth, the damage to the NFO's credibility loomed large.[42]

Many who marketed livestock prior to the existence of the NFO also resented the organization's charges, believing that they had been trying to get the highest price for farmers also. They believed that prices were largely determined by supply and demand and not subject to the manipulations that the NFO charged. The Farm Bureau, which was well established in the Midwest, feared the "loss of control to NFO" and attempted to defend organizations "under attack by NFO." The Farmers Union, which supported government farm programs and cooperative interests, also resisted the NFO's "do-it-yourself philosophy" and competed with the NFO for the activist farmers. NFU president James Patton noted the "engendering [of] hatred and animosity among farmers which the NFO has set in motion in just two or three years" and regretted the "violence [which] expos[ed] the economic flank of the American farm family to renewed and more aggressive attack from its traditional enemies." The NFU and allies such as Senator Hubert Humphrey worried when the National Catholic Rural Life Conference supported the NFO, fearing any move that bolstered the legitimacy of the NFO to the detriment of the traditional farm program. Perhaps the most unfortunate rivalry involved established cooperatives, which worked for the same goals as the NFO but viewed NFO strategy as "pie in the sky" and "economic fantasy." Many viewed NFO actions as disruptive of marketing arrangements that took decades to build and were especially damaging to smaller marketing cooperatives. The battles between the groups led to the prolonged and expensive "Midwest Milk Monopolization Case" and contributed to the beginning of federal probes into the cooperative enterprise, the first large-scale effort since the Capper-Volstead law passed in the 1920s.[43]

Despite these problems, the NFO story also indicates that greater farmer organization can, at times, strengthen farmers' bargaining efforts and increase returns to the farmer. The most promising period of NFO bargaining, it should be noted, occurred during a period when the packing sector was relatively unconcentrated and displayed signs of competitiveness, a coincidence that strengthens the argument in favor of including relative bargaining power as a factor in antitrust analysis. The NFO effort also stimulated greater awareness of market power issues among farmers and contributed to the development of various farm marketing techniques. Their success in engaging the alleged monopoly problem by improving farm marketing can also be seen in the growth of cooperatives in the postwar period.

6

FARMER COOPERATIVE MARKETING

James Patton, president of the NFU, believed that the monopoly problem underscored the need to promote organized marketing to build farmers' market power. The Farmers Union held that "cooperatives are potentially the farmer's best answer to the growth of corporate agribusiness, and have served as a countervailing force and an alternative in the marketplace." The National Council of Farmer Cooperatives agreed that cooperatives provided "a partial countervailing effect on big agribusiness firms." Some even went a step beyond cooperative marketing and into cooperative processing.[1]

The idea of farm bargaining power that the NFO promoted originated with the cooperative enterprise, an institution that dated back to 1620 in North America. In the post–World War II years the cooperative marketing approach persisted, building on previous experiences, especially the experiments of the Populists and state-encouraged cooperative building during the 1920s and 1930s. Previously localized cooperative enterprises expanded, federated, centralized, merged, and began to compete with private merchants, elevators, processors, and exporters in a significant way—by the 1970s seven cooperatives entered the Fortune 500. A Chicago businessman and a Michigan grain dealer feared such successes would lead to the coming of the "cooperative commonwealth."[2] By 1969–70, 2,539 cooperatives handled a net volume of grain of $3 billion, or about 32 percent of the market, compared to 40 percent handled by the four largest grain companies. By 1983, 2,300 cooperatives marketed 38 percent of the nation's corn and soybeans.[3]

Many historians have overlooked the growth of farm cooperatives when assessing the postwar American political economy. The stagnation of the New Deal in the late 1930s, the "end of reform," spelled the coming of the "corporate commonwealth" in the postwar years, according to many historians. Political gridlock after the 1938 elections, the need to prepare for the war, FDR's rapprochement with big business, and the emerging consensus around a "politics of growth" involving a modest social safety net and reliance on military Keynesianism, instead of a fundamental restructuring of the political economy, became the legacy of the "New Deal liberalism" that defined American politics for the next forty years. Such an interpretation, which usually focuses on big business and big unions, neglects the political economy of agriculture. Robert Griffith, one of the few who actually recognize agriculture as a necessary factor in this interpretation, fits it into the "corporate commonwealth" story by arguing that the defeat of the Brannan farm program in 1950 proves corporate influence over postwar agricultural policy. Left out of the analysis are non-commodity program efforts to aid farmers, such as the promotion of agricultural cooperatives. Another recent publication, by one of Robert Griffith's students, typifies the neglect of the farmers' efforts to organize themselves in the new economic order. Elizabeth Fones-Wolf's *Selling Free Enterprise* extends the "end of reform" argument by detailing the eclipse of the 1930s vision of "equal rights, industrial democracy, economic equality, and social justice," all due to the mounting pressures and protests of business in the postwar years. But again, while attempting to support the "corporate commonwealth" thesis, Fones-Wolf pays no attention to agriculture, one of the largest sectors of the American economy.[4]

The larger, more powerful postwar cooperatives offered an alternative to traditional market channels and increased farmer bargaining power and competition among food processors. As with the NFO, the cooperatives still faced many challenges and suffered many failures. But their emergence, persistence, and success offers further evidence that farmers could organize their marketing.[5]

Protests against the prevailing "rules of commerce" in the late nineteenth-century United States triggered the largest effort to organize cooperatives to date. Similar to the NFO's "blocking" of commodity sales in the 1960s, Texas cotton farmers in the 1880s started "bulking" their crops so they could "act together as a unit in the sale of their product." Promoted by the Texas Farmers' Alliance, the idea of cooperative commodity marketing spread with the Farmers' Alliance and became the "soul of the Populist faith," ultimately leading to the creation of "the world's first large-scale working class cooperative." Again, similar to the NFO's exporting program, the Farmers Alliance Exchange of Texas pooled farmers' cotton and sold it to buyers in England, France, and Germany. Other "large-scale," "broadly gauged cooperatives" formed in other agricultural regions to "combat" the "forces of monopoly." In 1890 the Kansas and Missouri Alliance, along with the Kansas Grange, formed the American Livestock Commission Company with twenty-five thousand dollars in member capital and within six months enjoyed forty thousand dollars in profits. Presaging another NFO tactic, farmers even met in Lawrence to consider holding their crops off the market, hoping for better prices.[6]

The cooperative movement of the late nineteenth century failed to organize farmers to the extent that agrarian leaders hoped. In the 1890s the cooperative emphasis of the farmer movement gave way to money issues and Populist politics and, by the end of the decade, declined further with rising agricultural prices. In the 1920s came a revival of cooperative efforts to organize farmers' marketing, making the decade "the greatest in the history of cooperative development." Led by the cooperative organizer Aaron Sapiro, farmers attempted to extrapolate the organizational success of the California orange and raisin growers to all parts of the agricultural marketing system. According to Sapiro, the organization of farmers into marketing cooperatives would allow them "to control flow of supply as to time, place, and quantity, so that [farmers had] something to say about the conditions that affect price values." By 1923 promotional efforts led to the formation of over sixty-six cooperative associations with an annual volume of $400 million dollars; in 1930 cooperatives marketed $2 billion worth of products annually.[7]

The state aided the cooperative movement. The first significant statutory help came with the Capper-Volstead Act. Prior to the passage of the act in 1922, many policy makers were wary of potential monopolistic practices among agricultural cooperatives, especially the milk producers. With the coming of the farm bloc in the early 1920s, however, bolstered by the agricultural depression of 1920–21, agricultural leaders used the act to secure antitrust exemptions for agricultural cooperatives. For at least two decades after passage the "Congress, the USDA, and the Department of Justice acted as if farmers had been granted a substantial, if not limitless, immunity." Secretary of Commerce and later President Herbert Hoover also aided the cooperative movement. He saw the cooperatives as the agricultural equivalent of industrial trade associations, which were at the center of his larger organizational vision of the American economy. His advocacy of the Capper-Volstead Act, the Cooperative Marketing Act of 1926, and the Agricultural Marketing Act of 1929 were all part of his strategy to rationalize agriculture, "one of the most disorganized and chaotic of all the natural resource industries," by ending "cutthroat competition, chronic instability," and farmers' "extreme individualism."[8]

The success of the "California plan" (as Sapiro called it) and the milk cooperatives could not be tailored to the needs of other commodities. Orange and milk cooperatives were successful at controlling their market because they took advantage of a tradition of cooperative arrangements, the availability of public information on output and shipment, the complexity of market entry, and the manageability of the surplus. Grain and cotton, on the other hand, were stricken with monitoring problems because of the vastness and scope of their production, larger and storable inventories (as opposed to perishables such as oranges and milk), and ease of entry into the market. As a result, cooperative wheat pools at their height never included more than 5 percent of total production. The failure of cooperatives in the tobacco, prune, and apricot market can be traced to the same organizational problems. Three-fourths of the cooperatives that were formed were small elevators, livestock shipping associations, and dairy creameries in the grain belt. As David Hamilton has noted, "they routinely sold to private handlers in the larger markets and refused to cooperate with one another to reduce costs. [and] . . . they were too small to improve the farmers' bargaining position or to make substantial gains in marketing efficiencies."[9]

Despite problems, Hoover supported cooperatives throughout the decade. As early as 1924 he sought federal legislation establishing a Federal Cooperative Marketing Board and a Federal Cooperative Marketing System to promote cooperative organization. Hoover emphasized the need to build the economic bargaining power of the farmer and encouraged the establishment of centralized commodity associations that could handle the task. He urged the three thousand grain elevators to form a Central Grain Marketing Association that could control 50 percent or more of a year's marketing of wheat and slow the pace of marketing when prices dropped to prevent a further downward spiral. From the beginning of his campaign, however, Hoover insisted that the membership in the cooperative enterprise must be voluntary, avoiding, as Hamilton has written of Hoover's view, "regimentation and bureaucracy, a loss of free will, the decline of self-government." As a past famine relief coordinator, Hoover took a dim view of "surplus control" measures that may have compensated for the increased productivity stemming from mechanization and biotechnology while they increased prices and aided cooperative organization. Many existing cooperatives were also resistant to Hoover's plans.[10]

After he was elected president, Hoover forged ahead with his plan for a Federal Farm Board that would coordinate a new cooperative marketing structure. In addition to providing loans to budding cooperative enterprises, the legislation creating the Farm Board authorized stabilization corporations that would buy and sell commodities to avoid large price fluctuations. The goals of the Farm Board were to develop cooperatives powerful enough to regulate farmers' production and to encourage marketing through centralized commodity associations that eliminated competition among cooperatives.[11]

Many who served on the board were not convinced that Hoover's plans would work, and many cooperatives resisted "amalgamation," preferring to remain independent of a national commodity association. Many joined only to take advantage of cheap credit—3.5 percent instead of 6–8 percent. In 1929 only fourteen hundred of the thirty-four hundred cooperative grain elevators joined their commodity association, known as the Farmers National Grain Corporation. And the private trade, which was in jeopardy of being displaced by the new marketing program, denounced the Farm Board's "socialistic tendencies" and its stabilization program. During a Chamber of Commerce meeting in Washington in 1930, the Minneapolis grain traders and their allies called for an end to loans to cooperatives and to stabilization efforts. Despite the pleas of the Farm Board, farmers did not curtail production—they planted the same amount in 1930 as they did in 1929, ignoring the Board's "Grow Less, Get More" campaign. In 1930 John A. Simpson was elected president of the National Farmers Union by defeating C. E. Cuff, a less radical leader who emphasized the building of cooperatives and served as president of the Farmers National Grain Corporation. Simpson, on the other hand, opposed Hoover's promotion of production controls, viewing them as detrimental to smaller farmers, and advocated legislation that fixed farm prices, further jeopardizing the cooperative-building experiment.[12]

From 1929 to 1932 the Farm Board loaned $360 million to cooperatives and continued to emphasize the importance of cooperative marketing. The amount of commodities marketed through a cooperative increased 15 percent from 1930–31 to 1931–1932, and more and more local cooperatives joined regional and national marketing associations. By 1932, seventeen hundred local cooperative elevators were members of the twenty-seven regional cooperatives affiliated with the Farmers National Grain Corporation, and the National Livestock Marketing Association was also growing. Much of this progress can be attributed to the generosity of the Farm Board, however, as many problems inhibiting cooperative organization persisted. Rivalries among cooperatives remained intense, cooperatives affiliated with the Farmers National continued to market through private channels that were more profitable, and too few farmers understood the workings of the organization or considered the national associations "farmer-controlled." Organizations such as the National Livestock Marketing Association, which did not rely on Farm Board funds to remain profitable, were more successful when they allowed greater autonomy to local affiliates, a deviation from Farm Board policy. Another cooperative affiliated with the Farmers National was Farmers Union Terminal Association, managed by M. W. Thatcher, then a solid supporter of President Hoover. With around 30 percent of the votes in the Farmers National Grain Corporation (based on the amount of stock held in the corporation), Thatcher was able to acquire Farm Board funds to cover his organization's losses and gain authority over his competitors who were members of the Farmers National but who had fewer votes, raising questions about the Farm Board's fostering "unprofitable, often autocratic institutions" and, in turn, its promotion of large-scale, centralized commodity associations in general.[13]

With the onset of the Great Depression, the building of marketing cooperatives as a remedy to the farm problem competed with several other legislative proposals and was overshadowed by the production control efforts that came with the New Deal (see chapter 7). Nevertheless, the cooperative movement persisted. As a result of the Farm Credit Act passed in 1933, the funds remaining in the Farm Board's cooperative loan fund were transferred to the newly created Central Bank for Cooperatives and twelve district banks for cooperatives, including banks in Minneapolis, Omaha, and Wichita. By the end of 1934 the system had loaned $68 million to cooperatives; in 1941 the annual amount loaned equaled $221 million, and by the end of World War II this increased to $407 million. In 1962, 60 percent of regional cooperatives' borrowed capital came from the Banks for Cooperatives. Unlike the Farm Board system, cooperatives that received funds from the Central Bank were not coerced into national commodity associations. The cooperative research agency created by the 1926 Cooperative Marketing Act was continued by the Farm Credit Administration's Cooperative Research and Service Division. Although some cooperative leaders feared that the federal farm program would displace their institutions,

Chester Davis, an AAA administrator, argued that "with controlled production, the worst pitfalls will be gone from the path of the cooperative movement."[14]

In addition to production control, the New Deal programs addressed several problems that previously slowed cooperative development. Access to capital, which severely handicapped the cooperatives of the Populist era, became less of a problem with the establishment of the Banks for Cooperatives. The banks also bolstered the financial position of cooperatives by improving management techniques, requiring financial statements, and conducting audits. Information and research generated by the agencies aided grain- and livestock-marketing cooperatives, and more cooperatives began to process, market, and advertise their own commodities. The political position of the cooperatives remained strong through the existence of the National Council of Farmer Cooperatives (which included most of the large cooperatives), the National Livestock Marketing Association, a federation of regional grain cooperatives, and the American Institute of Cooperation. State cooperative councils or associations were also powerful, especially in states such as Wisconsin, Nebraska, South Dakota, Minnesota, and Michigan. By 1948, twenty-eight councils existed in twenty-seven states (Wisconsin had two), promoting cooperatives, offering educational programs, and participating in legislative debates. Over one-half of the councils had budgets of $12,000–35,000, mostly financed by the dues of cooperative members. From 1933–34 to 1939–40 the volume of commodities marketed through cooperatives increased from $1.2 billion to $1.7 billion.[15]

In the grain-marketing field, the Central Bank for Cooperatives continued the Farm Board's previous promotion of the Farmers' National Grain Corporation. When the National dissolved in 1938, the component cooperatives were reestablished on a sound footing with help from the cooperative banks and with the leadership of the manager of the largest cooperative involved, M. W. Thatcher of the Farmers' Union Grain Terminal Association (GTA). In order to reinvigorate the local elevators that composed the GTA, Farm Security Administration loans were made to farmers so they could buy shares in cooperative elevators, a move that revived 177 elevators in midwestern and northwestern states. To add to the political power of the cooperative movement, specifically the power of the grain cooperatives, the National Federation of Grain Cooperatives was formed in 1939 with Thatcher as president. From 1940–41 to 1942–43 the regional grain cooperatives increased their grain handling from $104 billion to

$183 billion. In livestock, the Central Bank for Cooperatives aided the cooperatively organized Detroit Packing Company, which "served as an experimental laboratory or pilot plant for cooperative livestock packing, and much was learned on the problems of operating a plant of this type." And by 1943, twenty-four cooperative mills producing soybean oil and high-protein meal were running. After World War II, in order to highlight the power and development of the farmer cooperatives and take the "sting" out of Soviet charges of "Yankee imperialism," the great cooperative supporter Senator Arthur Capper (R, KS) urged Secretary of State George Marshall to route supplies to Europe through cooperatives.[16]

The growing success of the cooperative enterprise also generated opposition, especially from privately owned businesses forced to compete with cooperatives and their economic advantages. When tax rates soared during World War II, the tax-exempt status of farm cooperatives, conferred on them during the years of cooperative promotion during the 1920s, gave them an enormous competitive edge over their privately organized rivals. The Consumers Cooperative Association, now known as Farmland Industries, withheld patronage funds from its farmer members so it could finance manufacturing operations, including oil refineries for its consumer cooperative operations, which supplied farmers. "Factories are Free" was the slogan of the CCA's general manager, Howard Cowden. In 1943 Thatcher's GTA bought the St. Anthony and Dakota Elevator Company with its 135 elevators and 38 lumberyards, a move that "shocked the grain trade." The Arkansas Wholesale Grocers' Association protested to its members GTA's enormous start-up loan from the government, exemption from income taxes—of the $2 million in profits for 1943, $1.4 million would normally have gone to taxes—and failure to pay their fair share of the war effort. The result was the formation of the National Tax Equality Association (NTEA), funded by grain traders, oil and lumber manufacturers, and other business interests, including Standard Oil, Sears, U.S. Steel, Cargill, and Firestone, designed to publicize the problem of "unequal taxation" and seek legislative redress.[17]

An internal NTEA report documented the "many unequal competitive advantages [petroleum cooperatives] enjoyed over privately owned oil companies [which] made them formidable competitors," including "exemption from income and excess profits taxes and business regulatory laws, together with governmental credit and special privileges [which] enabled [them] to absorb many independent distributors and refiners." The NTEA cited fourteen cooperatives that earned $10.4 million in fiscal year

1942 and, under the tax rate of 63.4 percent, should have paid $6.5 million in taxes. Instead, the cooperatives were able to keep $7 million of their earnings for expansion and returned $3 million to their members in cash and stocks. With such financial advantages, the NTEA feared the coming of "super-cooperatives" producing and manufacturing oil and, ultimately, the coming of "international cooperative cartels to dominate world markets," especially after the formation of the International Cooperative Trading and Manufacturing Association in 1944. The Peavey grain company also protested the money that competitors such as Thatcher's GTA could reinvest in the expansion of elevators and other facilities. Recognizing that cooperative taxation was a political "hot potato," Peavey tried to build support for the NTEA by informing its employees that their job was at risk if the tax advantages continued and reminding them that as taxpayers they were being cheated.[18]

Abilene Flour Mills Company protested to Senator Capper that "it is well known that cooperatives do not return all profits to their members and that they do buy buildings and buy out other companies and corporations and expand because they do not have to pay income taxes like the corporations do." Their protest included a bulletin from the Arkansas Wholesale Grocers' Association which highlighted the "growing menace of COOPERATIVES" and called on businessmen to present the tax information to local Chambers of Commerce, Rotary Clubs, Lions Clubs, and other groups before what Vice-President Henry Wallace called the "bloodless revolution" ushered in the "cooperative commonwealth."[19]

The National Council of Farmer Cooperatives responded with the establishment of the National Association of Cooperatives, pledged to "correctly informing the American public as to the vital place of farmer cooperatives in the American system of private enterprise." Thatcher, who especially feared for the future of cooperatives after the death of FDR in 1945, emphasized the "choice between the marketing of the farmers' products and the furnishing of his supplies through co-operatives, with the saving going to the farmer, or the marketing and selling under private institutions, with the profits going to a few." Similarly, Congressman Henry Jackson (D, WA) attacked the NTEA as an "undercover outfit" that was trying to drive cooperatives out of business.[20]

Despite the NTEA's campaign, public opposition to farmer cooperatives remained low and political support remained strong. In August 1945 *Fortune* concluded that the NTEA "has driven the various kinds of cooperatives into closer cooperation than if they had been left alone. Already the

NTEA seems to be doing more to promote the cooperative movement in the public eye than it has ever been able to do for itself." A poem used in the political battles illustrates the high ground enjoyed by the coops:

Why are they after farm Co-ops,
In organized packs night and day?
Because, they're the wolves of production,
And like wolves they devour and slay.

There is just one way to fight wolf packs,
And that is to fight in a band,
Demanding our rights to fair returns
On the wealth we produce from the land.[21]

The NTEA and other critiques were effectively countered by the political allies of the cooperatives. In 1948 the Minnesota Association of Cooperatives and Thatcher's GTA supported Minneapolis mayor Hubert Humphrey in the Democratic-Farmer-Labor Party (DFL) Senate primary, opposing the communist-backed candidate, whom they thought could be defeated by a Republican favorable to the NTEA's tax arguments. The cooperatives also targeted Minnesota Republican congressman Harold Knutson for defeat, a thirty-two-year incumbent who chaired the House Ways and Means Committee, which wrote tax law. Knutson placed the termination of the cooperatives' tax advantage high on his agenda after the 1946 election, alienating farm support in his rural district. He was also a target of the CIO, who had "pour[ed] money" into his district to help his opponent in 1946. In addition to defeating a rural Republican uninterested in labor issues, the CIO wanted to preserve its cooperative enterprises. The UAW, for example, had twenty retail stores in Michigan in 1948 that did $20 million in business, in addition to cooperative housing projects and credit unions. The NTEA pointed out that "CIO UAW president Walter Reuther has set half of all American business as the goal of the tax-dodging consumer cooperatives that he is promoting." As the 1948 elections neared, the cooperative tax issue was even more prominent as Senator John Williams (R, DE), representing the many corporations that took advantage of his state's less restrictive incorporation laws, offered the anti-cooperative tax amendment on the Senate floor, claiming it "constitute[d] the most important test of our domestic economic policy ever to face the U.S. Senate," one that would "determine the future of our system of free enterprise."[22]

Humphrey won the nomination and his race in the fall against an incumbent senator, Congressman Knutson was defeated, and Truman surprised the nation by winning the presidency by swinging the midwestern farm vote. Although normally a Republican stronghold, the heavy concentration of cooperatives hurt Republicans in the Midwest, who were seen to be pushing congressional investigations of cooperatives and as friendly to the NTEA. Hence, "in future years Republican legislators firmly turned their backs on proposals to extend the corporate income tax to cooperatives." A farm lobbyist later said that "an attack on cooperatives [was] akin to denouncing the flag." Also, at the time of Humphrey's nomination his assistant, Orville Freeman, graduated to the chairmanship of the DFL, beginning the career of another staunch political ally of cooperatives. Understanding these past political debts and underscoring the political importance of the cooperatives in the postwar years, Humphrey's first public address after he lost the presidential race to Richard Nixon came in December 1968 to the annual convention of the GTA. "GTA has been a Godsend to its members," he reminded the audience he had spoken to for nineteen consecutive years. "We must never permit anything to happen legislatively, or any other way, that will cripple the effectiveness of the great farm cooperatives."[23]

In September 1952, during the presidential campaign, NFU president James Patton met with candidate Dwight D. Eisenhower. Patton explained that farm service cooperatives couldn't make much money if they bought supplies from other companies and sold them to farmers, so they decided to manufacture their own farm supplies—Farmland building refineries, for example. He argued that "farmers' coopcratives require help to establish their own unassailable sources of supply—this help essentially consists in that they be permitted to accumulate the capital that such developments require, without its being taken away from them by punitive and unfair taxation instigated by their competitors. Such taxation can stop the cooperative movement dead in its tracks, and make it impossible to start new cooperatives or expand older ones." That same month in a speech in Omaha, Eisenhower made clear he understood: "Farmer co-operatives are an essential device for maintaining the independent family farm. We will not let them be endangered. We shall aid farmers to strengthen their own institutions."[24]

In 1958 the NCFC sought joint resolutions in Congress reaffirming "national policy to aid and encourage the establishment, operation, and growth of farmer cooperatives" as a way of deflecting potential changes in the tax laws. When the secretary of the treasury recommended changes in the tax law that would force cooperatives to pay income taxes, Thatcher sought the help of Vice-President Nixon and succeeded in killing the initiative, noting that "we licked the NTEA by the action that was taken by Mr. Nixon." In 1959 Thatcher was able to get Republican senators Karl Mundt (SD), Carl Curtis (NE), Frank Carlson (KS), Milton Young (ND), and Nixon to end the IRS practice of requiring retroactive payment of taxes on patronage funds not returned to cooperative members within seventy-five days after the close of the fiscal year. To maintain the cooperative system and its privileges Thatcher also relied heavily on Democrats such as Eugene McCarthy, Hubert Humphrey, George McGovern, and Orville Freeman, whom he called "my boys." At the time of his retirement, one newspaper noted that "perhaps the biggest of all Bill Thatcher's victories in battles for farm people was in the co-op tax fight of the 1950s."[25]

Cooperative advocates such as Senator Capper also promoted the creation of an agency in the USDA dedicated solely to cooperative research and development. After passage of the Farm Credit Act of 1953, the Cooperative Research and Service Division, formerly located in the Farm Credit Administration, became directly controlled by the secretary of agriculture within the USDA. Newly named the Farmer Cooperative Service, it carried out the Cooperative Marketing Act of 1926 and contained its own administrative structures, making the service "a self contained and complete organization." Among the service's five divisions was the Marketing Division, which supplied information to cooperatives based on commodity lines, including grain and livestock. Farm cooperatives could request information, market research, and technical assistance from the FCS to aid in the growth and efficiency of their institutions. In 1954, for example, the FCS responded to the request of a Midwest regional grain cooperative for a study of its grain-marketing and storage operations. After analyzing the terminal elevators and the 143 member country elevators, the FCS recommended a plan to "consolidate and modernize all farmer-owned grain marketing facilities and expand storage capacity," which they estimated would lower marketing costs 40 percent. In the 1960s the FCS, working with the Banks for Cooperatives, offered the market information and technical advice that launched the Producers Export Company, owned by regional cooperatives, which exported millions of bushels of grain and oilseeds to thirty different countries.[26]

In 1959 Senators Long of Louisiana and McCarthy of Minnesota introduced a bill to exempt cooperatives from section 7 of the Clayton Act, which severely limited mergers. The Farmers Union Jobbing Association supported the bill, arguing it would increase bargaining power for farmers, reduce duplication, and, if the mergers succeeded, would spread more wealth among more farmers, instead of concentrating "wealth in the hands of a few." Deputy Attorney General Lawrence Walsh argued that the bill "would grant blanket immunity from the antitrust laws" and urged caution. One supporter of cooperatives argued that the whole idea should be dropped so it didn't feed into the "coop monopoly" arguments of the NTEA. Perhaps the Senate's foremost antitruster, Estes Kefauver (D, TN), killed the bill because he opposed any measure that would weaken the antitrust laws. In 1964, echoing promises made by Senator Kennedy in 1960, President Johnson's farm message declared that "new legislation is needed to clarify the right of cooperatives to expand their operations by merger and acquisition. I shall shortly transmit to the Congress, also, legislation to provide additional credit facilities to permit rural cooperatives to assume additional responsibilities in the war to combat poverty."[27] As early as January 1965 the Farmer Cooperative Service was advocating rural development in the form of local processing and assembling cooperatives and encouraging existing cooperatives to aid in the planning and formation of new cooperatives. In 1966 the Office of Economic Opportunity granted $150 million to the National Cooperative League to organize cooperatives among the poor and $3 million to help seven hundred low-income farmers in Wisconsin start the Wisconsin Feeder Pig Marketing Coop.[28]

Another mechanism for strengthening farmer cooperatives was to extend to them, as a whole, the same marketing options presented individual farmers under the price support provisions of the farm program. In June 1961 the Commodity Credit Corporation announced it would recognize cooperatives as producers in the soybean price support program. The executive secretary of the National Federation of Grain Cooperatives knew that the aid to cooperatives would infuriate private grain traders, who would make Secretary Freeman's "telephone and telegraph wires hum like hornets" as part of their "highly organized, well-financed and frenzied campaigns of opposition." They did, and their pleas were ignored. In June 1973 the Nixon administration announced plans to make grain cooperatives eligible for price support loans on wheat, barley, corn, sorghum, rye and flax. Although the plan was shelved because of internal squabbles in the USDA, the idea was revived by Secretary Bob Bergland in 1977. Cooperatives coalesced behind the plan, which, as the Nebraska Cooperative Council argued, aided in the "orderly marketing" of farm commodities. Private grain traders, unable to take advantage of the price floors guaranteed to grain cooperatives by the farm program, vehemently opposed the change. The president of the Salina Board of Trade told private traders that if the change was made "many elevator operators feel that most of us can plan on being out of business sometime within the next five years." The grain trade challenged Bergland's decision in court, but lost.[29]

Food manufacturing, which was accused of various monopoly sins and the subject of the National Food Marketing Commission's massive probe in the 1960s, was a sector many cooperative supporters encouraged cooperatives to enter. A study of the market activities of 100 of the 150 largest farmer cooperatives in 1977 revealed that of the total value of shipments, 38 percent stemmed from food manufacturing (most of the remainder stemmed from cooperatives supplying farm production inputs such as feed, fuel, and chemicals). Of the 20 largest cooperatives in the study, 4 were listed in meatpacking, 3 in flour and grain mill products, and 6 in soybean oil milling. In the latter sector, for example, involving 65 firms and a top eight concentration ratio of 73 percent, cooperatives held two of the top eight positions. Cooperatives held thirty-five different top-four positions in twenty-four different food product classes. The range of processing activities carried on by the 100 cooperatives in the study was high, averaging around eight different product classes; for the 20 largest cooperatives, the average was around sixteen. In 1975 the large dairy cooperative Land O'Lakes declared that "we have an objective of becoming a total agricultural/food company. A unique status, a position that many are coming to recognize as the means by which cooperatives will generate maximum returns for the farmers who own, who control and direct the destiny of their cooperatives." And diversify they did. The more complicated and multifaceted the processing, however, the lower the rate of cooperative involvement. While milling wheat into flour remained relatively easy, manufacturing the flour into bread and crackers and marketing them remained relatively difficult. Similar problems affected such sectors as cereal, baking, beer, and liquor, which were deeper into the processing stage, involved a much smaller amount of raw farm commodities, and required advanced marketing and advertising techniques. The largest cooperatives, like the largest food manufacturers, tended to be the cooperatives who participated in the sectors with the highest value added. There remains some question about how much the cooperative processors directly engaged the monopoly problem, given a "strong affinity by cooperatives for relatively unconcentrated product classes," and ones with little product differentiation. The same year as the cooperative study, the one hundred largest food manufacturers accounted for 55 percent of the value added in food manufacturing. The occasion for entry seems unmistakable, however, as the several success stories indicate.[30]

Several grain cooperatives also began exporting their supplies, increasing their share of port elevator capacity from 10 percent in 1968 to 21 percent by 1980. As export sales increased, so did their efforts to foster free trade and undermine protectionist measures in Congress. Kenneth Naden, executive director of the NCFC, asked Secretary of Commerce Maurice Stans in 1969 not to consider trade restrictions on textiles "'in isolation' from other trade matters," fearing they could lead to restrictions on imports of American agricultural commodities. The NCFC, the Farmers' Union, the Farm Bureau, the Grange, and other farm groups all lobbied President Nixon to be wary of trade negotiations that could "jeopardize American foreign markets" by inviting "more retaliatory barriers to our exports." Such stands pitted the liberal farm groups against would-be political allies, such as factory workers in the South. Senator Everett Dirksen (R, IL), who favored restrictions on textile imports, noted that "there are many who attribute the conditions in Appalachia and the necessity for substantial expenditure of federal funds to the loss as a result of imports of textile products." It allied the farmer advocates with groups they often denounced and ridiculed, such as the U.S. Chamber of Commerce, but they proved a powerful influence in promoting free trade and blunting the protectionist urge of the 1970s and 1980s.[31]

Several examples offer the best evidence of cooperative success. The Farmers' Union Marketing and Processing Association (FUMPA), for one, started in the 1930s and continued into the postwar years, gaining increased attention in the 1960s' New Frontier "atmosphere . . . favor[ing] . . . rapid expansion and growth of cooperatives" and farmer bargaining power. In 1967 the FUMPA sold nearly $5 million dollars worth of its members' cattle and their by-products, the latter being processed in three FUMPA rendering plants. Sales increased to $28 million in 1973, serving as a "very substantial income producer" according to the National Farmers Union president. In 1976 FUMPA marketed over twenty-two thousand cattle and sixty thousand hogs through its facilities in St. Paul and West Fargo, which contributed to the $1.55 million in profits it made that year, and ended the year with nearly $9.5 million in assets.[32] The Montana Farmers Union also attempted to establish the Montana Livestock Cooperative in 1975 with plans for three packing plants, hoping to cut the cost of transporting cattle to other states for slaughter, but the plan failed to generate enough interest among livestock growers.[33]

The most important example of cooperative success, according to the agricultural historian Gilbert Fite, is Farmland Industries Inc. Farmland started during the cooperative revival of the 1920s as the Union Oil Company and expanded into a vast enterprise, mainly due to its activities as a farm supply cooperative. In the 1930s the leadership argued that the payment of refunds should be slowed and the money used to fund the expansion of facilities, creating "tremendous power" that would "sweep on down stream to enrich the farms and homes of cooperative members." In 1939 the cooperative built an oil refinery in Phillipsburg, Kansas, to help supply the fuel to be sold to farmers. In 1943 CCA teamed with Central Cooperative Wholesale of Superior, Wisconsin, Midland Cooperative Wholesale of Minneapolis, Farmers Union Central Exchange of St. Paul, and the Farmers Union State Exchange of Omaha—collectively known as the National Cooperative Refinery Association—and bought another refinery in Texas with help from the Central Bank for Cooperatives. Refinery operations expanded further and by the end of World War II CCA operated 369 oil wells, which supplied 18 percent of the oil needed for their refineries. CCA's further diversification included lumber mills, canneries, alfalfa and potato dehydration facilities, feed mills, a milking machine factory, fertilizer plants, grocery stores, a farm implement plant, a phosphate plant, and a $16 million nitrogen plant—all while the CCA newspaper headlines included "Cartels Cast Lustful Eyes upon the World" and "Vice-President Wallace Blasts Monopoly." In 1957 CCA was 327th on the Fortune 500 list.[34]

About this time, after prompting from a cooperative member from Mt. Pleasant, Iowa, CCA moved into farm marketing and processing. In 1958, in Eagle Grove and Ida Grove, Iowa, CCA began boar testing as a first step toward farmers' developing quality hogs for market. The next year CCA bought the Crawford County Packing Company in Denison, Iowa, and created the subsidiary Farmbest Inc. to operate the plant as a cooperative marketing association; in 1963 another hog plant was opened in Iowa Falls. Farmbest did not generate the kind of income for CCA that the oil operations did because it operated in an extremely competitive market. By 1967, after CCA had changed its name to Farmland Industries Inc., the hog operation included sixteen buying stations, slaughtering, cutting, ham and bacon processing, and the marketing of Farm-King, Farmbest, and Country Manor. By 1966 the total investment in the operation came close to $6 million, a large part of which was owed to the Omaha Bank for Cooperatives. In 1975 Farmland constructed another $18 million hog slaughtering facility in Crete, Nebraska.[35]

In 1965 Farmland dedicated a cattle-slaughtering plant in Garden City, Kansas, known as the Producers Packing Company. A plaque mounted on the new facility read: "This plant is dedicated to the building of a strong livestock industry based upon producer and feeder participation in the processing and marketing of meat." In three years the independent cooperative, then with a kill capacity of seven hundred daily, merged with Farmland. As with many small packers during the period, the facility lost money every year from 1965 to 1968, including $620,000 in 1967. In 1970 the Farmland Board of Directors voted to create Farmland Foods Inc., a new corporation to handle the meat business. A man who had been with Farmland for six years and Swift for twenty-five years before that became president. Farmland wanted farmers to be able control their hogs from "first oink to the dining table," and by 1973 Farmland processed 10 percent of the hogs raised in Iowa. Patronage funds paid to farmers in 1971 totaled $1.22 for hogs and $2.48 for cattle. In some years, however, no patronage refunds could be paid at all. Although the number of livestock cooperatives declined from 538 in 1950–51 to 510 in 1969–70 (in large part owing to fewer farmers and cooperative mergers) and the percentage of the livestock business conducted through cooperatives declined, the volume of business increased from $1.3 billion to $2.1 billion and cooperatives still sold over 10 percent of livestock.[36]

Another great cooperative success was the Farmers' Union Grain Terminal Association, or GTA, led by its three-pack-a-day general manager M. W. Thatcher. When the National Grain Corporation unraveled in the 1930s, Thatcher made sure the GTA emerged as a strong regional cooperative with a great deal of support from FDR. Thatcher recalled: "I had been in business about six months when the bank for co-ops, the Central Bank in Washington, turned me down. I went to the President. He told the governor of the FCA [Farm Credit Administration, which oversaw the cooperative banks] that he wanted to see him, and to bring his resignation with him." Thatcher received a loan of $300,000, compared to the $30,000 put up by the Farmers Union Central Exchange. By 1939, ninety-three local cooperative elevators were organized or reorganized by GTA, and by 1944 the cooperative represented 150,000 grain producers and marketed over $100 million in grain.[37]

In 1943, to formalize a bargaining process for farmers, Thatcher proposed a National Agricultural Relations Act to give agriculture the "same right to self-government and adequate returns that Labor was granted," presaging the large-scale campaign for bargaining power that the NFO would stimulate in the next decade. Thatcher and the NFO sounded much alike: "The natural law of supply and demand has been amended by man-made law. The so-called 'free market,' by which is meant simply a market governed by nothing but the law of supply and demand, has passed out of the picture.... American business grew big and powerful under the laws granted it in our legislated economy.... At the turn of the century the American laborer began to share in a few of the simpler benefits of our legislated economy.... Far-sighted farm leaders also have recognized, down through the years, that farmers must organize or be lost." Thatcher organized Farmers' Union locals and GTA members to support the measure, but the effort was largely overshadowed by the debates over fixed versus flexible parity in the postwar Congresses. Not until the introduction of the Mondale bargaining bill in 1968 did Congress consider another full-scale legislative proposal to reorganize the process of farm bargaining, in part owing to a probargaining resolution adopted at the 1966 GTA annual meeting.[38]

In the meantime, GTA grew into a powerful regional marketing and processing cooperative, largely centered in Minnesota, the Dakotas, and Montana, which also made the organization the "big muscle in the National Farmers Union." In 1941 GTA built terminals in Superior, Wisconsin—"the tallest, fastest-operating elevator in the world, not to be matched for many years"—and Lewistown and Shelby, Montana; in 1942 GTA bought the Amber Milling Company in Rush City, Minnesota, and soon ground 20 percent of the nation's annual durum wheat harvest. The next year, GTA acquired St. Anthony and Dakota Elevator Company and soon leased four more elevators with a storage capacity of 4.5 million bushels. By 1950 GTA membership numbered 615 local and line elevators with a total capacity of 28 million bushels, plus terminal elevators holding another 19 million bushels. In twelve years GTA used $22 million in "savings" (profits) for this expansion and paid back $3 million to farmers and their local cooperatives. It also paid $1 million in "educational funds" to the Farmers Union. In 1958 GTA bought McCabe and Company (Ben McCabe was the president of the International Elevator Company and president of the NTEA, as well as the head of McCabe and Company) and its elevators and feed mills in fifty-seven locations for $4.8 million. It soon bought another thirty-seven elevators, the largest soybean-processing facility in the country, and Minnesota Linseed Company, which processed food and sold paint and varnish.[39]

In 1960 GTA not only purchased the Archer Daniels Midland elevator line but also moved further into processing with the acquisition of the Honeymead soybean-processing plant in Mankato, Minnesota. GTA bought Honeymead from Dwayne Andreas and his brother. Andreas, an "Iowa farm boy" from a 160-acre farm near Lisbon, Iowa, who quit college after one year to run the local elevator, had built several feed plants and processing operations, which he sold to Cargill in 1945 for $1.5 million. He then worked for Cargill for seven years and led its move into processing—by 1952 they had eight processing plants. He then started Honeymead, processing and exporting on his own, selling the largest soybean operation in the world to GTA in 1960 for $6 million in cash, and staying on with them in a management position. Invoking a cooperative principle, Andreas argued that "it makes sense for farmers to own their processing and marketing business." With the help of Andreas, GTA bought the Minnesota Linseed Oil Company in Minneapolis in 1961 for $3.6 million and the Froedtert Malt Corporation in Milwaukee in 1965 for $10.5 million. After the acquisition of Froedtert, Andreas recommended that GTA make no further moves into processing and instead move into exporting. Although considered by some Thatcher's heir, Andreas resigned from GTA and formed Interoceanic Corporation with his brother, ultimately buying a large portion of the shares of Archer Daniels Midland. Thatcher concluded that "the large and complex business advances we have made during the past five years at GTA would have been impossible without Mr. Andreas' knowledge and assistance." When the grain business sagged in the mid-1960s, the processing divisions "were the only part of the [GTA] operation that made any money."[40]

In 1975 GTA earned $29 million on $800 million in grain sales, $425 million in processed grain products, and $40 million in building materials, ending the year with assets of $273 million. In relation to assets, GTA earned about the same amount as Cargill. Just before the Carter grain embargo, at the annual convention, GTA announced that its 780 locally owned cooperatives marketed nearly 400 million bushels of grain, earning nearly $30 million on assets of $530 million.[41]

The equivalent of GTA in the southern part of the grain belt was Far-Mar-Co. In 1968 four regional grain cooperatives—Equity Union Grain Exchange of Lincoln, Nebraska, Farmers Cooperative Commission Company of Hutchinson, Kansas, Cooperative Marketing Association of Kansas City, and Westcentral Cooperative Grain Company of Omaha—merged to form the Farmers Marketing Company, or Far-Mar-Co Inc. The merger came with the help of officials with the Banks for Cooperatives and after ratification by the members of the cooperatives. An indication of the growing clout of the organization came after Bill Thatcher stepped down as president of the National Federation of Grain Cooperatives after twenty-nine years and the post was filled by Jimmie Dean, the general manager of the newly formed Far-Mar-Co.[42]

Far-Mar-Co provided accounting services to its member cooperatives through its data-processing service. It also moved into research, patenting a process for the fractionation of wheat and developing Starea, a livestock feed concentrate, and processed soybeans and wheat. In 1971 Far-Mar-Co handled 217 million bushels of grain and in 1972 considered joining GTA in buying out Archer Daniels Midland and all of its processing facilities, which could have been supplied with GTA–Far-Mar-Co grain. The same year, Far-Mar-Co considered a merger with ADM and Farmland, pulling together their meatpacking, grain-processing, and farm supply activities, the "ideal for maximum benefit to the midwest farmer," but the merger did not work out. At the time of its final merger with Farmland in 1977, Far-Mar-Co handled 350 million bushels, served six hundred local elevators in nine states, owned sixteen terminal elevators, owned or leased one thousand jumbo hopper cars, and controlled part of the Farmers Export Company.[43]

The merger with Farmland also brought the opportunity for more efficient marketing and expansion. As a subsidiary of Farmland, Far-Mar-Co designed a "centralized grain marketing system with significant improvements in control and supervision," in which grain transactions were reviewed by high-level managers. Far-Mar-Co also drew on the research capabilities of Farmland when seeking avenues for expansion. The Economic and Market Research Division of Farmland, for example, produced detailed feasibility studies of potential grain terminals in Salina, Kansas; Enid, Oklahoma; and O'Neill, Osmond, Merna, and Superior, Nebraska. Researchers paid close attention to the market potential for these sites, weighing the available grain supply, storage and handling capabilities, the "competitive conditions" in the area, and the potential building, operating, and transportation costs. In the case of Superior, the study noted the good soil of the Republican River bottom, several U.S. highways in the area, Superior's three railroad carriers—Burlington Northern, Santa Fe, and Missouri Pacific—and the several cooperative and noncooperative elevators in the area around Superior. The study estimated the return on investment for a new Far-Mar-Co terminal handling 20, 25, and 30 percent of the Superior-area grain to be 17, 19, and 27 percent respectively, "returns that are normally considered acceptable for this type of investment."[44]

In addition to mergers and expansions, Far-Mar-Co also sought to organize farmers as part of the growing emphasis on "marketing agreements," or arrangements in which producers legally bound themselves to an organization that would market the farmers' crops. With advice from the Farmer Cooperative Service, Far-Mar-Co started the Producers Marketing System, or Promark, which marketed farmers' grain and paid them their share of the profits. Far-Mar-Co's general manager believed that "we should be able to expand the market for producers because of our strategic position: having the wheat in a known position; being able to do long-range planning for facilities, for finances and for transportation." The Promark program also emphasized "market intelligence" and market strategy guided by information from "processors, importers, exporters, contacts in foreign countries, private advisory groups, trade associations and government reports." In 1976, its first year of operation, Promark sold the wheat of fifteen thousand grain belt wheat farmers, and in 1978 Promark marketed over 22 million bushels of wheat, more than 60 percent of which was exported to places such as Brazil, Peru, Sudan, the Soviet Union, Pakistan, Portugal, Poland, Japan, Bangladesh, Israel, West Germany, the United Kingdom, Morocco, China, and Mexico.[45]

The experience with Promark seems to underscore the complexity and the intense level of competition in the export sector. A 1979 report on the Promark program concluded that "Murphy's Law has been at work since the inception of Promark four years ago.... The list of problems seems endless: transportation tie-ups, high interest rates, Farmers Export Elevator explosion, litigation, unpredictable markets, by-law changes at your cooperative, mounds of paperwork and government 'red tape' for both the cooperatives and Promark." Although the report concluded that Promark "in most cases has out performed the typical producer's buy/sell marketing decisions," many farmers began to lose faith in the meager results. In 1979 Farmland's Economic and Market Research Division conducted a survey of farmers who dropped out of the Promark program and found that one-half "found price received to be lower than price obtained through own selling activities." Again highlighting the independence that made collective action for farmers complicated, 20 percent indicated that they "disliked others making decisions for them."[46]

As Dwayne Andreas left GTA, he tried to arrange a deal between GTA and ADM and between GTA and Garnac Grain Company to share facilities for exporting. When this failed, GTA joined Far-Mar-Co in becoming part of Farmers Export Company, which comprised nine regional cooperatives when it started in 1966. Farmers Export Company built a five-million-bushel export facility in Ama, Louisiana, in 1969, "open[ing] the way to year-around exports for midwest cooperatives." The group planned another terminal for Galveston for more efficient movement of grain to foreign markets.[47]

The Farmland–Far-Mar-Co merger highlighted an important postwar cooperative trend, using mergers to grow larger and economically stronger "in order that they might meet the challenge of vertical integration among their competitors." Farmland illustrates the trend: in 1966 nine north-central Illinois cooperatives affiliated with Farmland; in 1967 they added the Minnesota Farm Bureau Service Company and its 14 cooperatives; and in 1968 they added the Southern Farm Supply Association and its 83 cooperatives. Thus, by 1970, 2,085 cooperatives were affiliated with Farmland. The Farmland–Far-Mar-Co merger in 1977 brought together the largest supply cooperative and the largest grain-marketing cooperative, generating sales in 1977 of over $3 billion and moving Farmland to seventy-eight on *Fortune*'s list. Other examples included the grain belt cooperatives Illinois Grain Corporation and FS Services, which merged to form Growmark Inc. in 1980, with $2 billion in sales, and grew further with the acquisition of St. Louis Grain Corporation. Land O'Lakes acquired the Spencer Beef packing operation, merged with Midland Cooperatives, and launched a joint venture with CENEX (formerly Farmers Union Central Exchange), which had merged with the Western Farmers Association. In the early 1980s GTA merged with North Pacific Grain Growers, creating Harvest States Cooperatives, with an annual grain handle of 500 million bushels, revenue of $3 billion, food sales, and the processing of beans, wheat, barley, and sunflowers. The press release claimed that "Harvest States will be the world's largest cooperative engaging in grain marketing, processing and farm supplies, and the only one with export facilities to the Great Lakes, the Gulf of Mexico, and the Pacific Ocean."[48]

As some older cooperatives merged, some newer cooperatives were formed. In August 1980 farmers living in and around Waseca County, Minnesota, gathered in the machine shed at Mert Hildebrandt's farm to talk about starting a corn-processing cooperative. Anticipating greater demand for high-fructose corn syrup because companies such as Coca-Cola and PepsiCo were moving away from using refined sugar in their drinks, they formed a board of directors and chose Mankato, Minnesota, as the place for a new corn sweetener plant, "which would be the first farmer-owned wet milling plant in the nation!" They originally sought to avoid red tape by seeking private funds in lieu of money from the Banks for Cooperatives. This effort was displaced by another to build a corn-processing cooperative to produce ethanol in Marshall, Minnesota, an effort that received greater support from the Banks for Cooperatives. The Marshall effort led to the formation of Minnesota Corn Processors, which processed the corn of nearly twenty-five hundred farmers in Minnesota, Iowa, South Dakota, and Nebraska by 1993. In the 1990s there is talk of the "'value-added' co-op movement" and "co-op fever," particularly with the end of the grain belt government support programs after passage of the Freedom to Farm Act.[49]

While growing stronger economically, the political and legal position of farmer cooperatives diminished slightly starting in the 1970s. During the Carter administration the Farmer Cooperative Service's "agency status" was ended and its functions merged into the Economics, Statistics and Cooperatives Service (ESCS). Randall Torgerson, former administrator of FCS and now administrator of the Rural Business-Cooperative Service, informed Farmland officials that FCS would continue in the same capacity, but that the "net effect" of the change de-emphasized cooperatives. Without agency status, the FCS could also be more easily undermined if a less supportive secretary of agriculture took over. The executive secretary of the Wisconsin Federation of Cooperatives argued that after such a change "only the very largest and most sophisticated, multi-state cooperatives (which are probably least in need of concerted governmental aid), would be sophisticated enough to be able to permeate the beaurocratic [sic] monster that is USDA to utilize these cooperative personal," undermining the cooperative efforts of "small family farmers and beginning co-ops." The National Council of Farmer Cooperatives as a whole pressed the USDA to maintain a large-scale research and policy analysis component to aid cooperative formation; Senator Nelson cited the example of how USDA technical assistance assisted in the planning for a large cooperative grain export terminal elevator in Milwaukee, an effort that helped the "little guys" in agriculture.[50]

Kenneth Farrell, administrator of the ESCS, ordered a committee to review the programs, purposes, and future of the cooperative unit, particularly in light of growing "resource constraints." The committee finished its work early in 1980 and noted its preference for independent agency status for the cooperative unit and argued that the unit's "raison d'être" should be spreading knowledge about cooperatives while focusing on activities with "maximum leverage effects." The committee also criticized the unit for being a "jack of all trades, and a master of none of them" and encouraged a narrower scope of research, service, and advisory committees and more specific criteria for the selection of projects to aid. Resistance led to the recreation of the Farmer Cooperative Agency in the 1980s.[51]

The NTEA also continued to dog the cooperatives. They won a small victory with the passage the Revenue Act of 1962, which required cooperatives to pay 20 percent of refunds in the form of cash, not stocks or certificates to be redeemed later, a change many farmers found hard to oppose. In 1969 the NTEA lobbied to raise the requirement to 50 percent but failed, as did their efforts to work with senators such as Abraham Ribicoff (D, CT) to amend the Tax Reform Act of 1970. The tax issue received less attention after the election of Jimmy Carter and the coming of what the NTEA called "a liberal, co-op oriented Administration," anchored by Vice-President Mondale and Minnesota farmer–turned–Secretary of Agriculture Bob Bergland. Bergland: "My father would come after me with a gun if I ever sold any grain any way except through our local co-op" (a fertilizer cooperative also supplied the Carter peanut farm). After Congressman Ed Mezvinsky (D. IA) called for an investigation into the role cooperatives played in food price inflation in 1976, he was defeated in the fall election, partly because of his position on cooperatives.[52]

Of greater concern to cooperative leaders in the 1970s was the revival of questions about their antitrust status and the potential repeal of their Capper-Volstead exemption. Instead of legislators discussing a wider cooperative exemption from the antitrust laws—the much-debated exemption from any merger restrictions in the 1950s and 1960s, for example—the discussion turned to ending the cooperatives' Capper-Volstead exemption from the antitrust laws, what the NTEA called the connection between "cooperative monopolies and their immunity from antitrust laws." The controversy started, as did so many others, with Richard Nixon. During the presidential campaign of 1972, the CREEP's fundraising apparatus included Associated Milk Producers Inc. (AMPI), which made contributions to the campaign in exchange for increases in dairy price supports, setting off a spate of public and press scrutiny of cooperatives. The inflation of food prices also generated political pressure to suppress marketing activities that increased prices, despite farm groups' efforts to divert the attention with calls for another large-scale study of food pricing to update the National Commission on Food Marketing's report of the mid-1960s. The large-scale mergers such as Farmland and Far-Mar-Co also generated public attention.[53]

In 1973, when testifying to Congress about the problem of inflation, the head of the Department of Justice's Antitrust Division suggested that agricultural cooperatives needed examination. Two years later another assistant attorney general suggested that cooperative dairy mergers that resulted in $10 million in business in any market area should be prohibited. Senator Humphrey feared that such comments meant "the Antitrust Division of Justice may be preparing a new 'offensive' against the [Capper-Volstead] Act." In 1974 Secretary Butz told the Republican Antitrust Task Force that he agreed that the special antitrust exemption afforded cooperatives needed to be investigated, and in 1975 Butz told students at Harvard that GTA was too dominant in states such as North Dakota: "They have gobbled up all the private elevators along the Milwaukee and Northwestern Railroads." Butz went into the lions' den in 1976, telling the convention of the NCFC that some cooperatives had abused marketing orders for "monopolistic price enhancement." The FTC's Bureau of Competition would add to the public scrutiny by calling attention to the milk industry's "huge regional supercooperatives." The NFU and other farm groups saw the DOJ, the USDA, and the FTC as "engaged in politically oriented witch-hunt aimed at unfairly discrediting farmers cooperatives."[54]

The most focused effort to change the cooperatives' antitrust status came with the formation of the National Commission for the Review of the Antitrust Laws and Procedures in 1977. The commission concluded that cooperatives should continue to be allowed to merge, but only as long as "no substantial lessening of competition" resulted. Although not directly expressed, the recommendation contradicted a century of cooperative efforts that were designed to limit disorganization and competition among farmers and increase their collective market power.[55]

The commission's second recommendation involved the section of the Capper-Volstead Act that prohibited farm marketing cooperatives from abusing the market power granted by the legislation. Specifically, the law forbade "undue price enhancement" and granted the secretary of agriculture the power to investigate such charges and issue cease-and-desist orders. From the passage of Capper-Volstead in 1922 to 1979, however, only five investigations were conducted, and the secretary never found that a cooperative violated the provision. Some suggested that the enforcement of the "undue enhancement" provision be transferred to the Department of Justice, the FTC, or another newly created agency designed to monitor cooperative behavior. Secretary Bergland and other farm advocates explained how undermining cooperative law might "weaken competition" by affording greater relative market power to large food firms and other buyers and opposed any legislative changes. Kenneth Naden, executive director of the NCFC, agreed: "The antitrust commission has dealt the family farm system a low blow by attacking cooperatives, the farmer's main alternative to weak bargaining power." The NFU launched a "Don't Antitrust Our Co-ops Campaign."[56]

The antitrust debate highlighted some of the continuing dilemmas of the farmer cooperative enterprise. While farmers presented the best case and exerted the greatest political pressure for antitrust action, they in turn asked for legislative codification of inherently anticompetitive practices for themselves. In the 1970s, for example, they constantly fought any weakening of the protections and exemptions from the antitrust laws that they held. In the middle of the decade some legislators tried to divest the largest eighteen oil companies, and such a political effort would normally have been vigorously supported by farm groups—in the 1950s cooperative leaders such as Jerry Voorhis of the Cooperative League were attacking the "biggest oil money" for helping spread the views of the NTEA. But some farm groups blinked. The Minnesota Association of Cooperatives, for example, thought cooperatives such as Farmland and CENEX might lose their pipelines if the fury for divestiture spilled over into their areas and onto their enterprises. When the NTEA attacked the cooperatives' tax advantages and antitrust exemptions as anticompetitive, the cooperatives charged in turn that the NTEA was destroying competition by harming cooperatives. Thurman Arnold, whom the Cooperative League hired to outline a potential antitrust suit against NTEA, argued that NTEA lobbying to end the cooperatives' tax break constituted a "conspiracy to restrain, harass, and destroy the competition afforded by the cooperatives." In 1979 the NFU urged all members of Congress to support a bill by Congressman Mark Andrews (R, ND) to prohibit the FTC from ever using any tax funds to investigate anticompetitive practices of farmer cooperatives. Such awkward reasoning and favoritism diminished the political goodwill toward cooperatives.[57]

The growing economic power of cooperatives that generated calls for antitrust inquiries also generated resentment among some farmers. Victor Ray, a researcher for the NFU, noted the "alienation of [cooperative] patrons . . . a general feeling that the coop is impersonal, disinterested, and remote" because many technical decisions needed to be made by management. "The characteristics of cooperatives are becoming less distinguishable from those of large corporations." The minutes of a Farmland meeting recorded the perceived farmer attitude: "The strongest feelings from producers is that they are not represented by the bureaucracy of farm organizations and agricultural businesses. The majority of farmers feel that in their bigness, they have become so bureaucratic in their leadership that they have lost their relating ability to the country," a criticism also leveled at the large-scale cooperatives Hoover's Farm Board tried to build. In 1974 the Agribusiness Accountability Project released the study *Who's Minding the Coop?*, claiming cooperatives were getting too large, entering areas unfamiliar to farmers, and losing their democratic nature. Even some officials of the NFU, longtime defenders of the cooperative enterprise, feared the growing size of cooperatives and the power of their managers and predicted they might replace the general farm organizations.[58]

Cooperative leaders, on the other hand, claimed that bigness and market power were needed to compete with agribusiness. An agricultural economist at Texas A&M criticized the cooperatives for an absence of economic power, using the word "fragmentation" to describe the cooperative grain marketing sector. Joseph Knapp and Ronald Knutson, both heads of the FCS in the postwar years, constantly urged cooperatives to merge, become stronger economic forces, reduce intercooperative competition, and hire expert managers. Willard Cochrane told Secretary Freeman in the 1960s that to gain market power in commodities such as hogs, wheat, corn, and beans an organization must be large-scale, and not "local in character."[59]

Despite continuing antitrust questions and the dilemma of building market power at the possible expense of their democratic nature, many postwar farmer cooperatives succeeded. In the process, they offered farmers another marketing alternative that could be used to circumvent the monopoly problem. Like the NFO, they coordinated farmer marketing and served as a powerful organizational tool for farmers. The degree of their involvement in farm marketing and the ambitiousness of many of their efforts is a prime indicator that farmers were not always hapless victims of corporate power in the post–World War II years.

7

THE STATE AND AGRICULTURAL ORGANIZATION

> We need immediate action on this farm problem. Some of the young fellows who fought for this country can't make a living now & the older farmers are next. I am asking you "Is this right?" LLOYD ZUBROD, FARMER, IONIA, IOWA

The organizational problems that limited the effectiveness of the NFO and the cooperatives explain the origins of the federal farm program in the 1930s. One of the lessons of the Farm Board experience, it seemed to many observers, was the inability of farmers to organize to control their production. Farmers thus turned to organizing through the state, using the federal government to restrict output as they feared private firms in concentrated markets were already doing. "Politics," as Colin Gordon argues, "is largely about the method and degree of coercion used to restrict individual interest and the acceptance or legitimacy of public interest and private restraint." Many farmers in the early 1930s accepted state coercion in order to effectively restrict production, raise prices, and overcome their organizational problems. Throughout the grain belt, according to John Mark Hansen, "rural lawmakers who opposed organized agriculture, already a scarce breed, were pushed even further toward extinction." In the 1950s such prominent politicians as George McGovern of South Dakota emerged in the grain belt, continuing to argue the case for "a farm price stabilization program, given the unorganized pattern of American farm producers."[1] While the state proved to be a useful organizational tool at times, its unpredictability and coercive power also angered and alienated many farmers. The resulting ambiguity and the variety of farmer opinions limited the ability of the state to broker between interests in a way favorable to farmers. As the depopulation of rural America accelerated in the postwar years the willingness of the state to even attempt to broker proagricultural legislation dwindled further.

THE POLITICS OF THE FARM PROGRAM

During the summer of 1930, when the Farm Board was trying to build powerful marketing cooperatives, a first-term Republican congressman from rural southern Minnesota introduced a production control bill. Victor Christgau, the first member of Congress to hold a graduate degree in agricultural economics (from the University of Minnesota in his case), thought information from government experts could help farmers adjust their production to demand. He understood the potential resistance of farmers to national controls and planning, however, and emphasized that "we are still essentially a democratic country, living under a democratic system of government. We cannot dictate to the farmers what they shall produce, but we can lead them along certain lines and supply them with information." Farmers needed to be like U.S. Steel, the argument went, better able to adjust supply to demand and therefore better able to turn a profit. When the Farm Board collapsed and farm prices continued their plunge, plans such as Christgau's were tempting. In September 1932 in Topeka, Kansas, FDR endorsed the approach of Christgau and his supporters, increasing the odds of the major farm groups agreeing to production controls, and blamed the Republicans for the farm problem (Hoover refused to endorse production controls), defining farm politics and policy for the next forty years.[2]

Launching the Agricultural Adjustment Act's production control program was chaotic, but the program partially accomplished its goals. In the summer of 1933 the program drew criticism for plowing up cotton and tobacco and committing "pig infanticide" when millions were ill-clad and ill-fed. Secretary of Agriculture Henry Wallace and AAA administrator George Peek battled over the issue of production control almost immediately, Peek resisting "secret plans" for "regimented production," part of what Theodore Saloutos called a "virtual state of civil war" within the AAA. In a 1935 "purge," Wallace was able to eliminate many within the USDA who opposed production controls and favored other policy alternatives. Then in the first week of 1936 the Supreme Court struck down the AAA as unconstitutional. After an interlude with the Soil Conservation and Domestic Allotment Act of 1936, the farm program was reorganized in the AAA of 1938. Most importantly to the three-fourths of farmers who didn't lose their farms during the Great Depression, the various versions of the farm program helped to double cash receipts on the farm between 1932 and 1937.³

Production control and USDA management took a variety of forms and, unlike the National Recovery Administration's disastrous attempts at industrial planning, could rely on the USDA's research and information-gathering institutions in addition to civil servants and academic experts. In addition to the production restrictions of the early 1930s, the farm program involved marketing loans and quotas. The 1938 AAA, for example, made available marketing loans to wheat farmers if they agreed to an overall production quota in a referendum. If they approved the quota, they were eligible for a nonrecourse loan on their wheat, which supported the overall price level at somewhere between 52–75 percent of parity, depending on where the secretary of agriculture set it based on the estimates of USDA economists. If prices fell below the loan rate, farmers could forfeit their wheat to the government's Commodity Credit Corporation. If prices rose above the loan rate, farmers could sell their grain and repay the loan. As the price support functions became more deeply entrenched in the late 1930s, especially the non recourse loan, farm state politicians were tempted to maintain high support levels while weakening production restrictions. When World War II started, production restrictions were lifted; price support levels were fixed at 90 percent of parity for dozens of commodities and were to be continued by law at 90 percent of parity until two years after hostilities ceased.⁴

When the war ended it was unclear how the government would proceed with farm policy. Many feared the return of excessive production and depressed farm prices, but others feared excessive food shortages and global famine. As the wartime price support program was ending in 1948, the major parties involved in farm policy coalesced around the idea of returning to the farm program of the late 1930s, when price supports were "flexible." Many believed that support prices should move between 60 and 90 percent of parity, allowing for adjustments by the secretary during periods of surplus or shortage. President Truman, President James Patton of the Farmers Union, the Farm Bureau, and many important congressional leaders supported the flexible parity legislation. During the summer of 1948, however, when slackening postwar world demand became noticeable in farm prices, many Democrats abandoned the flexible support program and went on the attack. Although President Harry Truman earlier supported the flexible system, he blamed the Republicans for the problems in the farm program during the 1948 campaign. Echoing FDR's Topeka speech, Truman told an audience in Dexter, Iowa, that Republicans had "stuck a pitchfork in the farmer's back" and wanted "a return to the Wall Street economic dictatorship." After trailing Dewey in Iowa by close to thirty points during the summer, Truman went on to win the state and the election in November. His surprise victory meant he would need to offer a new program for farmers in the spring.[5]

One of the few administration officials who hadn't packed his bags in anticipation of Truman's defeat in 1948 was Secretary of Agriculture Charles Brannan, who echoed the president's harsh words when he spoke to farm audiences that fall, accusing Tom Dewey and other Republicans of attacking price supports. Brannan followed up his campaign criticism with a plan for aiding the farmers in the spring of 1949. Instead of continuing price supports based on pre–World War I parity prices, Brannan wanted to gear the farm program to the income level of farmers during World War II. Brannan wanted to allow the prices of perishable commodities such as hogs and cattle to fluctuate with the market, but proposed to offer farmers government payments if prices didn't maintain farmers' income at the designated level. If the government payments became too expensive, the secretary of agriculture was given the authority to impose production controls to reduce supply and increase prices. Nonperishable commodities like wheat and corn were also to be pegged to the new income measure but were to be supported using a system similar to that of the 1938 AAA, without compensatory payments. In the summer of 1949 Democrats met in Des Moines to plan their political strategy for using the Brannan plan, believing that if they used it as an issue in the 1950 campaigns they could pick up twenty House and four Senate seats. The unknown costs of such a program, the similarity between compensatory payments and "welfare," caps on payments, and objections to the high support levels all contributed to the plan's demise. The policy stalemate in Congress led to the decision to continue wartime high-support policies for another year. With the coming of the Korean War in 1950 supports were again fixed at 90 percent parity and were to be continued until two years after the war. Truman campaigned in Illinois in 1950, hoping to support the Brannan plan by supporting the reelection of his Senate majority leader, Scott Lucas. But the state that had elected Adlai Stevenson governor and Paul Douglas senator in 1948 turned out its majority leader in favor of former congressman Everett Dirksen in 1950. Albert Loveland, who won the Iowa Democratic Senate primary by touting the Brannan Plan, was also defeated that fall, effectively ending the short career of the Brannan Plan.[6]

In 1952 the same Governor Stevenson went to Fort Dodge, Iowa, this time as the Democratic presidential nominee, and told the crowd that his opponent Dwight Eisenhower actually favored flexible price supports. Campaigning for Stevenson in Fargo, Truman told his audience, "You'd better 'look out neighbor.' The last time you had a Republican administration, farm mortgages were being foreclosed so fast you couldn't count them," again sounding like FDR in Topeka in 1932. Eisenhower's response came at the National Plowing Contest in Kasson, Minnesota, where he promised to support the farm law currently "on the books." To some this meant 90 percent parity and to others this meant flexible supports, because the wartime support levels were to expire in 1954. Then came the "bitter controversy" over the Eisenhower administration's proposal to switch to flexible price supports in 1954, a move designed to check what Ezra Taft Benson called a "rapid drift toward a regimented agriculture." After passage of the administration bill the Democrats recaptured both houses of Congress in the fall. When they passed bills to restore high price supports, Eisenhower responded with vetoes, establishing the basic outlines of the farm policy gridlock of the 1950s. "The nub of our present problem," Benson argued, "is unrealistic support prices and futile attempts to control production."[7]

As high supports and minimal production control persisted, government stocks of commodities and storage costs grew. Wheat farmers could have forgone the harvest of 1954 entirely, since more than a year's supply of wheat had already been forfeited to the government through the price support program, nearly 900 million bushels. The fixed minimum national allotment for wheat was 55 million acres (down from 78 million from 1945 to 1949), but the annual needs of American consumers could have been provided on 19 million acres. Despite the disparity between supply and demand, Benson was still under pressure from Congress to lift allotment restrictions and to allow the planting of other crops on the diverted acres. "Members of Congress," Benson's biographers wrote, "were happy to take credit for USDA funds going into their district but were quick to disassociate themselves from government-imposed restraints."[8]

The same year that Benson became secretary of agriculture, George McGovern stepped down from his post as a history professor at Dakota Wesleyan University and filled the post of executive secretary of the South Dakota Democratic Party, supplying much of the ammunition that kept Benson in the crossfire for the next eight years and providing an apt synecdoche for the farm politics of the 1950s. The most powerful agricultural organization in South Dakota in the postwar period, and McGovern's most important political ally, was the Farmers' Union. For the South Dakota Farmers' Union to advance its agenda, however, it needed a political party, and the Democrats in South Dakota were in complete disarray, a common condition in the traditionally Republican heartland. When McGovern took over as head of the Democratic Party in 1953, Republicans outnumbered Democrats in the state legislature 108 to 2, and a Democrat had not been a member of the congressional delegation in nearly twenty years, a party dominance common in the Republican states of the grain belt since their settlement. Even by 1961 a Rapid City Democrat still argued that "neither Christ or Martin Luther, seeking election to the U.S. Senate, Congress or as Governor, on the Democratic ticket in this state could expect to be elected even if the opposition party had a slate of unknown candidates and SD Democrats were oozing unity and practicing political togetherness."[9]

Even absent Christ or Luther, many South Dakotans were willing to vote for Democrats based on their pledge to solve the farm problem. In the 1930s, 56 percent of the population lived on the state's 83,000 farms. But owing to the depression, changes in farming practices, and the dislocations associated with World War II, many families were leaving the farm, and as many saw it, the flexible price support policies of Secretary Benson hastened the exodus. The South Dakota Farmers' Union claimed credit for helping to stop the attempt to establish flexible supports in 1948 and denounced Eisenhower's repudiation of the "Golden Promise" of 90 percent parity prices made in Kasson. In March 1953, only a few months into Benson's eight-year term as secretary of agriculture, the president of the Miner County Farmers' Union in South Dakota wrote to President Eisenhower and requested he ask the new secretary to resign. By the following year the number of farms in South Dakota dropped to 62,500, and with every additional farm sale, grain belt politicians and groups such as the NFU criticized Benson more and more harshly.[10]

McGovern made the most of the farming issue. He criticized Eisenhower's failure to honor the "Golden Promise" and blamed "eastern Republicans" and "eastern capital which directs the GOP" for undermining price supports. In a speech to the North Dakota State Democratic Convention in 1954, McGovern cited a poll indicating that 52 percent of farmers would vote for Adlai Stevenson instead of Eisenhower a year after the administration was in office and noted the growing number of former Eisenhower supporters in South Dakota who had joined the "Never Again

Club." McGovern also reported on the new slogan that was "sweeping the farm belt"—"Vote Democratic; the farm you save may be your own"— and candidly admitted that any Democratic gains in South Dakota would be as a result of Benson's unpopularity.[11]

In 1956 McGovern stepped down as executive secretary of the party to run for one of the state's two congressional seats; the most important issue of the campaign was agriculture, as it was in many grain belt political contests during these years. His opponent in the 1956 contest was the five-term incumbent, Harold Lovre. Senator Karl Mundt, the dean of the South Dakota congressional delegation, warned Vice-President Nixon in late 1955 that the "anti-Benson sentiment in the farm belt is something which is just pretty serious. They not only do not like his farm program, but now they don't like the man." In January 1956 Lovre went to Benson with two Iowa congressmen and begged for price supports for hogs. One of the Iowans pounded on Benson's desk, telling the secretary that "if you don't put supports under hogs, not one of us will return to Congress next year." "For decades the Tall Corn State had been rock-ribbed Republican territory," according to Edward and Frederick Schapsmeier, "and now stalwart GOP congressmen were running scared." A formal poll taken in neighboring Wisconsin had shown that the percentage of farmers who thought Benson was doing a poor job had grown from 13 percent in July 1953 to 55 percent in January 1956, the month that McGovern launched his candidacy. When Joe McCarthy died the following year, three-time gubernatorial loser William Proxmire ran for the Senate seat and "geared his entire campaign for election on an anti-Benson theme." In his 1956 race Proxmire only won 30 percent of the rural vote, but with his anti-Benson campaign in 1957 he received 70 percent and won the race. In 1959 Quentin Burdick was elected to the Senate in North Dakota with the slogan "Beat Benson with Burdick."[12]

McGovern worked closely with the Farmers' Union throughout the campaign, using the group's twenty-three thousand members to raise money and organize voters, and agreed to have all his farm speeches cleared by the editor of the *South Dakota Union Farmer*, whom he would take to Washington as his administrative assistant.[13] McGovern was also helped by the newly formed NFO, which flooded Washington congressional offices with calls for Benson's "hide," worked with the Farmers Union to advance their then similar goal of parity price supports, and ultimately formed chapters in fifty-six of South Dakota's sixty-seven counties to advance their agenda. McGovern won the contest by eleven thousand votes, mostly through very strong showings in rural areas, and Eisenhower's plurality in the state dropped from 69 percent in 1952 to 58 percent in 1956. Unsurprisingly, the first congressional speech of McGovern's career addressed the farm problem, and within a few months he called for the head of Secretary Benson, as did many chapters of the NFU and NFO in South Dakota and around the grain belt.[14]

Republicans kept Benson out of South Dakota during the fall campaigns, and the reason became even more clear. After the elections the December issue of *Harper's* included an article entitled "The Country Slickers Take Us Again," criticizing "our pampered tyrant, the American Farmer," charging that farmers sold their votes to the highest bidder, and claiming that the "average" Iowa farmer had a minimum of two cars, usually including a new Buick, Oldsmobile, or Cadillac. The following month a letter to the editor called the article "excellent"; it was signed by Ezra Taft Benson. In the storm of criticism that followed, Senator Francis Case (R, SD) said that if the letter was true, then Benson had ended his "usefulness as Secretary of Agriculture." Benson claimed that an assistant had signed the letter and apologized for pulling such a "boner." Case then called for suspension of the assistant for being "disloyal to agriculture and disloyal to the Secretary"; the South Dakota Farmers' Union and the GTA made the most of the gaff. The incident further exposed the divisions in the Republican Party over the farm issue as the Republican lieutenant governor criticized Senator Case for criticizing Benson's bumbling in press releases just as Benson was visiting South Dakota and asked the senator for "minimum publicity" in the future. Benson's popularity sank so low that later in the year he was showered with eggs at the National Corn-Husking Contest in Sioux Falls. The *Huron Daily Plainsman* chided Benson for calling the incident "un-American" and cited the American revolution as evidence that "there is nothing more in the American tradition than the heaving of groceries"; one DeSmet preacher thought it possible that a monument to the "embattled farmer" might be built in Sioux Falls in "emulation of that at Concord Bridge." Knowing he needed to shore up his chances for reelection in 1960, Senator Mundt released press statements underscoring the distinctions between his farm policies and Benson's when the secretary visited South Dakota. Some Republican politicians such as Mundt took the "anti-Benson bait," as the *Sioux City Journal* called it, and openly called for Benson's removal.[15]

After using his first congressional speech to bash Benson's farm policies, McGovern proceeded to offer legislative alternatives. In the farm bill debates of 1957 McGovern offered an amendment to circumvent the secretary of agriculture's pricing discretion and fix the price of corn at 90 percent of parity. The amendment failed by four votes. In 1958 McGovern forwarded a "comprehensive" bill to deal with all of agriculture, making all farm commodities eligible for marketing orders that would strictly regulate agricultural production, a proposal similar to the Brannan plan. The bill didn't pass, but McGovern considered it and the proposal's original author, who then worked for the NFU, staples of the postwar "liberal cause"; some predicted that the plan would become part of the 1960 Democratic platform. The restrictions on farmers that would have been necessary for such a plan to work were heavy. One report indicated that "under strict regulations of a scarcity program" hog production would need to be cut 30 percent, wheat 55 percent, and eggs 30–40 percent in order to raise prices to the desired level.[16]

In 1958 McGovern's opponent was Governor Joe Foss. Foss tried unsuccessfully to convince Eisenhower to change his farm policies; the other congressman from South Dakota, a Republican, wrote to Eisenhower's chief of staff, Sherman Adams, and argued that a good man like Foss was not worth sacrificing on the "Benson altar." Senator Mundt instructed Vice-President Nixon not to "let anybody deceive you into thinking that Ezra Benson is any bargain in the farm belt. A blind man in a basement at midnight should be able to tap his cane and find that Benson's great unpopularity in the farm precincts cost us a bunch of Congressional seats." McGovern made the connection clear to his audiences by arguing that Foss "would just be one more vote for Benson and the big corporation-style farm advocates." Foss also lost ground by criticizing the Democrats' politicization of the "farm problem" as a campaign tactic and by waffling at the state corn-picking contest by saying that he was neither for nor against Benson. Mundt reported to one constituent that even the most prosperous farmers were disgusted enough with Benson to vote against Foss. Three months before the election McGovern led Foss among farmers 68 percent to 32 percent.[17]

After McGovern won the 1958 race he immediately began thinking of running for the Senate against Karl Mundt in 1960, a race one columnist called the year's "most important" senatorial race. McGovern danced with who brung him, hammering away on the farm issue. At a fifty-dollar-a-plate fund-raiser in Washington to kick off the campaign, which was attended by such presidential prospects as Senators Kennedy, Humphrey, and Johnson, who would all echo his views on farm policy, McGovern declared that "if Nixon is elected, with men like Mundt who support him . . . the family farm is doomed as an institution and corporate agriculture will sweep the country." McGovern attacked Mundt's "surrender" to Benson and linked the two together by citing the secretary's compliments of the senator and the fact that one of Mundt's former aides went to work for Benson. Mundt responded by citing the long list of favors he had done for agriculture since he was elected to the Senate in 1948. Mundt's reminder to voters that he had asked Benson to resign as early as 1953, coupled with the strength of his record on agriculture and his political organization, seemed to be enough to prevent a blowout in heavily agricultural precincts, and he hung on to defeat McGovern.[18]

President-elect Kennedy thought McGovern fit the bill of the postwar liberal intellectual "best and brightest" and wanted to work him into the administration. McGovern's friends in the Farmers' Union, Robert Kennedy, and Arthur Schlesinger Jr. all advocated McGovern for the secretary of agriculture post, but he was passed over in favor of Governor Orville Freeman of Minnesota, who lost his bid for a fourth term in 1960. Instead, McGovern was chosen to head Food for Peace, the surplus disposal program initiated under the auspices of Public Law 480. The law was another response to depressed farm prices and had enormous support in South Dakota, partially since it originated with the work of Senators Mundt and Case—one newspaper editor called it like being "for motherhood and against sin." McGovern used the publicity from Food for Peace and the aggravation in farm areas, especially that exhibited during the first large-scale NFO holding action, to seek and win the other South Dakota Senate seat in 1962.[19]

An issue in the 1962 campaign was the farm program that Secretary Freeman promoted for the Kennedy administration. Part of the new program proposed giving the secretary of agriculture extraordinary powers over the writing of the farm program, eliminating the congressional tendency dating from at least 1938 to bid up price supports to budget-busting levels. In the new proposal, the secretary would instead seek advice from commodity committees, two-thirds of which would be selected from the committees making up the USDA's Agricultural and Conservation Service and one-third would be selected by the secretary from the different farm organizations. After consulting with the commodity committees, the secretary would design a commodity's farm program, then seek approval from the president and from farmers through a referendum (Congress had sixty days to veto a particular program before the program became law).

The plan assumed, as Don Hadwiger and Ross Talbot wrote, that "Congress would recognize its inability to enact production controls, and would restrict itself to setting the broad guidelines within which the executive branch would actually write the farm programs." The assumption was flawed—excluding Congress from the policy-making process was not popular in congressional committees (unsurprisingly), and the idea was dropped.[20]

If the administrative change had been approved, the new secretary would have emphasized greater production control, an approach designed to overcome the policy gridlock of the Eisenhower years, which perpetuated a system of high price supports with minimal production controls. Instead of lowering price supports in order to reduce production, as Benson had done, the Kennedy program emphasized the need for production controls. In the final months of the 1960 campaign Kennedy told the audience of the National Plowing Contest in Sioux Falls what was coming, a farm program requiring "sacrifice and discipline" from farmers. Similar to Gordon's definition of politics, Kennedy argued that "men agree among themselves to limit their unrestricted 'freedom' in some field in order to achieve some other goal that is highly valued.... to circumscribe to some degree complete freedom to act in one field, to achieve a highly prized and generally accepted goal is, I repeat, the act of rational and civilized men." As Kennedy's farm policy advisor Willard Cochrane saw it, for Benson's plan to work, "support prices would have to go much lower and stay there a long time to bring production back into line with demand," and "no political party in modern times could live through the social and economic unrest that the above action would entail." Robert Clodius told the American Farm Economic Association in 1959 that "it should be evident that the policy of flexing support prices downward and moving toward the free market has not reduced production enough, increased consumption, reduced surplus stocks, and solved the farm problem." To avoid the costs of the farm program and to increase farm prices, Cochrane proposed a program of strict scarcity, setting levels of output in a system of "supply management" that would result in the proper prices, just like a "public utility," he admitted. Cochrane admitted that such a plan "entailed some relinquishment of farmers' freedom to make all their own production decisions." He believed what was "lacking [was] the courage on the part of politicians, farm leaders, farmers themselves and farm economists who serve as advisors to place in operation any one, or combination, of the alternatives that has the capacity, but which hurts someone, to cope with the excess capacity problem." Since new powers to design farm policy were not conferred upon the secretary, the supply management approach would need to be pursued through other legislation.[21]

Additional controls were not popular in Congress and received lukewarm support from farm groups. When the Kennedy administration submitted the Food and Agriculture Act of 1962 to Congress, a measure that included mandatory supply management programs for wheat, corn, and other commodities, only the wheat program survived. Since the legislation was not passed until September 1962 and planning for the 1963 wheat crop was underway, the program would not begin until 1964. The farmer referendum could be held, however, and it was set for May 1963. If farmers voted yes they would be accepting stricter mandatory production controls and higher support prices in lieu of the existing program's weaker restrictions and less generous supports. The 55-million-acre minimum national allotment for wheat had been abolished by Congress in order to reduce production to levels that would boost prices to the levels sought. Although this program survived Congress, it was voted down by farmers in the referendum, the first time a referendum failed to pass since 1939.[22]

The scramble for a new wheat program involved calls for new legislation based on the "emergency" feed grains bill of 1961, the temporary legislation passed when the supply management approach couldn't be prepared in time. Instead of forcing farmers to participate in a mandatory program, the 1961 law was voluntary. If farmers volunteered to reduce production by a certain amount, they would be eligible for price supports and payments on the amount of land taken out of production. Unlike the supply management plan, which relied on dramatic reductions in supply to boost farm prices, the voluntary plan relied on direct government payments to farmers, which cost the federal government more. The supply management proposal for corn was dropped in favor of continuing the voluntary program, and wheat producers thought the same could be done for wheat after the defeat of the referendum.[23]

After winning his 1962 Senate race McGovern was in a position to offer a new proposal, what came to be known as the voluntary wheat certificate plan. Because it was voluntary like the corn legislation, the new proposal did not require a producer referendum. Farmers could voluntarily reduce their output in exchange for acreage diversion payments and wheat certificates, which added another seventy cents per bushel to the price of wheat. Aided by lobbying from groups such as the NFO and GTA, the new wheat legislation passed in 1964. The following year the passage of the Food and

Agriculture Act of 1965 "marked the end of vociferous, and often inflammatory, congressional debates" dating back to the Brannan plan. The 1965 legislation embodied the voluntary approach, partially obviating the control and regimentation issue, and was more bipartisan, receiving the votes, for example, of all the Republicans from Kansas, South Dakota, and Minnesota. In 1968 a coalition of farm groups coalesced around the task of extending the act, which was to expire in 1969. Given the impending presidential election, the act was extended for one year to give the new administration some say in the final legislation and was again extended in the Agricultural Act of 1970 and the Agriculture and Consumer Protection Act of 1973. In the 1973 act, government payments were designed to make up the difference between the market price and the "target price" and came to be known as "deficiency payments." Farm policy debates of this period largely involved budgetary matters, which were not a concern during periods of high farm prices and low government payments such as the early 1970s. The Food and Agricultural Act of 1977, passed during the Carter administration, maintained this system but increased loan rates and target prices slightly.[24]

THE STATE AND THE BROKERING OF FARM INTERESTS

For some, organizing through the state seemed an obvious choice. According to an NFU official, price support programs give "the farmer the same kind of pricing machinery, in a general way, that every manufacturing industry enjoys." An Iowa farmer said in 1960 that "I don't believe the natural adjustment processes of the free market will ever allow farmers to achieve equity in the exchange of their products for the products of modern corporate industry or modern organized labor. We should use the market for the functions that it can perform well. But for the immediate future I do not see how we can avoid fairly drastic measures to bring our exploding production under control." The Grange, Farmers Union, GTA, and National Association of Wheat Growers agreed, and supported the 1963 wheat referendum. Before it publicly doubted and criticized the farm program, the NFO also conceded that "agriculture must either be regimented or it must discipline itself if it is to exist on an economic equality with other segments of our economy."[25]

Willard Cochrane believed that farmers "must come to accept production and marketing controls in the same way that they do driving on the right hand side of the road, paying their taxes, and sending their children to school. For the supply control approach cannot succeed unless the overwhelming majority of commercial farmers approve and accept it." Organizational efforts that weren't national, like the federal farm program, would not work in his view. In the cases of hogs, wheat, corn, and soybeans, Cochrane argued, "organizations designed to achieve market power through management of supplies in those commodities cannot be local in character. An organization having the purpose of achieving market power for wheat must include most or all of the thousands of producers involved, hence be nation wide in scope. And when an organization must be nation wide in scope involving thousands of producers we have not as yet found any alternative to government action." He further explained that the success of local cooperatives in California, for example, came from the localized nature of their markets and from using "their *political power* to win enabling legislation for putting into effect Marketing Orders, which if passed in referendum were binding on all producers in the area with respect to the control provisions involved." "Given governmental sanction," as Bruce Gardner notes, "a cartel of very many members becomes a plausible proposition. Indeed, the flue-cured tobacco program . . . is a fine example of a cartel, even though it contains many thousands of independent producers. The government, as manager of the cartel, each year sets a production goal, allocates it among producers, enforces a limit on sales of each member, controls entry absolutely, and allows no one to buy or sell the product outside the cartel's channels. OPEC should have such clout over its members!" Such views stimulated the various efforts of farm groups and farm politicians to formally extend marketing orders to all farm commodities. Congressmen McGovern and W. R. Poage, for example, pushed "nationwide marketing orders" in their Family Farm Income Act of 1960, and marketing orders were title 2 of Senator Mondale's farm bargaining legislation.[26]

References to the alleged organization and collusion of business were often made—some even thought farmers should have been given an NRA code of competition in the 1930s like industry enjoyed. The *St. Louis Globe Democrat* criticized free market advocates because the "complete return to the kind of laissez-faire 'free market'" that they supported was one "under which no American industry has operated for generations."[27] When some criticized the farm program for its degree of control and regimentation, farmer advocates such as M. W. Thatcher, mocked the regimentation argument ("I'm trying to find one farmer that's 'regimented.'") and made the comparison to the oil industry: "If you have restriction of acreage, then that's 'regimentation.' Well, the oil companies—they have the thing done right. I was working in Washington and lobbying for farmers when Senator Tom Connally of Texas brought in the 'Hot Oil Bill,'" which "provided for the complete regimentation of oil . . . that's good business for the oil companies, but if you regulate by quota how much wheat I sell off my farm, then that's 'regimentation,' that's like Soviet Russia."[28]

Acquiescence to the controls needed for the farm program to function was far from unanimous among farmers, however, and it was not popular among other groups with political power, such as consumers and organized labor. During the 1938 elections, after the establishment of what would be the core postwar farm program, "GOP candidates fanned agrarian resentment against the strict new production controls of the 1938 AAA," defeating New Deal senators in Wisconsin, Kansas, and Ohio and reclaiming twelve grain belt congressional seats. In western Illinois in 1938, as Lynnita Sommer has written, "much of the talk over the fence post and in the coffee shops centered on the government's authority to collect fifteen cents a bushel on all corn sold or fed above marketing quotas, if quotas were invoked." The talk led to the formation of the Corn Belt Liberty League and a meeting of over fifteen hundred farmers in McDonough County in April to protest controls. At a subsequent meeting a Madison, Wisconsin, man, unpersuaded by Thatcher-like mockery, warned that "the whole setup of the crop-control bill is designed to establish a dictatorship over agriculture. Every day the New Deal bites off a little of the freedom of the American people. Unless this is checked we will be in the same boat with Russia." After a trip around the Midwest in the 1930s, Rose Wilder Lane wrote that she would "vote for anybody—Hoover, Harding, Al Capone—who [would] stop the New Deal." Such resistance was part of a broader attack on the New Deal, as one historian recorded, by those who feared the "dead hand of government" and the "bureaucratic, socialistic, spendthrift schemes, which were shackling the energies and undermining the confidence of liberty-loving Americans."[29]

One objection was simply the politicization of prices, or making prices subject to the whims of powerful secretaries of agriculture. When Benson became secretary, for example, Thatcher alleged that the secretary favored western sheep and wool producers, because they were often Republicans and Benson was from Idaho and a Mormon, to the detriment of more Democratic wheat growers. Benson's Soil Bank program was also seen as pro–corn belt, "designed above all to benefit Republican farmers in the Corn Belt." The policy "stiffed the diehard Democrats who grew cotton in the South and the steadfast Republicans who grew wheat on the Plains" in order the bring states such as Iowa back into the Republican column in 1956. Farmers who stayed out of the voluntary farm program in the 1960s also objected to the politicization of their prices. They were dependent on market prices, which were held down by the government's decision to release excess stocks in order to induce more farmers to participate in the program. Some objected to the Kennedy bill's provision "whereby the farmers are threatened unless they vote yes at the referendum that the government will dump something like 200 million bushels from their stocks onto the market and thereby depress the market way down."[30]

Often the criticism of the farm program reflected an ideological opposition to economic controls and distrust of government economic management and a preference for a market-driven economy, a viewpoint represented by Benson:

Free enterprise: the right to venture—to choose
Private property: the right to own
A market economy: the right to exchange.[31]

During debate of the Brannan plan, a Kansas farmer said he did "not need . . . [or] want anyone to tell [him] how many acres [he could] plant. [T]his is supposed to be a free contry [sic] let us leave it that way." The 1952 Republican platform reflected such views and attacked the Truman administration for "seeking to destroy the farmer's freedom" and attacked the "Brannan plan which aims to control the farmer and to socialize agriculture."[32]

Nixon campaigned against Kennedy in 1960 on a farm platform pledging "freedom from controls." During the debate over the Kennedy legislation, the Farm Bureau president said the purpose of the Freeman bill is "cheap food produced by docile, licensed, and properly managed farmers," leading to a "regulated . . . controlled peasantry," and congressmen charged Kennedy with trying to create a "police state" and "regimented economy." The Farm Bureau feared that the wheat program would mean "a system of farming by government directives," calling it "the tightest, strictest, most complete control plan ever considered for a major farm commodity." The administration's 1962 bill, according to one Kansas farmer, was "only a smoke screen for their 1964 and later legislation to really put the screws to the farmer," and another called it "the greatest issue since the civil war and again liberty is the question." "The [Kennedy] administration's insistence on strict, mandatory production controls jeopardized the Democrats' recent gains in the Middle West," according to John Mark Hansen, and in political debates over farm policy, the "Republicans' most telling issue" was "coercion." In Kansas in 1962 freshman Republican congressman Robert Dole defeated the incumbent Democratic congressman Floyd Breeding, who was even chair of the House subcommittee on wheat.[33]

During the debate over the Kennedy program, Willard Cochrane admitted that more controls would be necessary. "It would not be a matter of encouragement; some [other commodities] would be forced in. If you had a control on hogs, for example, and none on eggs, growers would transfer their corn into the production of more poultry and eggs." Cochrane then thought the program would "yield prices in the market that have . . . been determined fair by some responsible agency." After his ambitious plans failed to be enacted, he conceded that "it is perfectly clear that farmers will not accept effective, mandatory controls." When Hubert Humphrey worked the farm areas for the Johnson ticket in 1964, he told audiences that the idea of compulsory production controls would be dropped.[34]

The opposition to government controls was an issue from the beginning of the farm program. In 1937 Gallup reported that 47 percent of farmers opposed a revival of the 1933 AAA. In 1972, 44 percent of farmers in a Dakota Farmer poll said that curtailed production was not the answer to low farm prices. When Mayor Richard Lugar of Indianapolis was campaigning for the Senate in 1974 he attacked the "liberal farm policies of Senators Bayh, McGovern and Humphrey [during which] the family farmer was driven off his farm by big brother in Washington." During the 1980s the Harkin-Gephardt farm bill proposed "mandatory production controls . . . [which] reformulated the demands of liberal farm groups from the 1960s and earlier" but failed to pass. Bruce Gardner noted in 1996 that 45 percent of farmers were opposed to commodity programs.[35]

As a result of the massive increases in agricultural productivity after World War II, the need for the production control that many farmers hated increased further. USDA experiment stations developed chemicals, genetics, plant varieties, and methods of cultivation that added to the surplus problem the USDA was trying to solve. The acreage allotments farmers were assigned kept producing more and more crops every year—in Iowa and Illinois thirty-five- and forty-bushel corn in the 1940s became eighty-bushel corn by the 1960s. As Dale Hathaway commented, "Land is only one input, and not a major one at that for most crops. There are other substitutes for land in crop production" such as "fertilizers, irrigation, improved seed, insecticide, etc." For example, the USDA predicted its 1961 feed grain program would decrease corn production 16-18 percent below the 1960 level. But corn yields increased six bushels an acre because of excellent weather and increased fertilizer use, and so production dropped only 11 percent. As John Kenneth Galbraith once said diplomatically, "This is an area where imperfection is in the nature of things." This situation came to be known as the "slippage problem," where farmers would divert the least productive acres and increase yields on the base acres through the many technological developments of the postwar years on top of farmers who opted out of the program and increased their acres of production. Although thousands of acres had been retired from production, as one economist pointed out, "the trouble is that farming know-how is increasing at such a rate that the resources that are left in American agriculture can regroup and go on producing a surplus." According to Gilbert Fite, "the cumulative effects of science and technology in agriculture defeated all of the policy makers' efforts to control price-depressing surpluses."[36]

The deep divisions among farmers about the degree and legitimacy of controls divided the farm lobby and complicated state efforts to broker an arrangement beneficial to farmers. Prior to the 1920s Congress had to listen to eighty-six hundred different farm groups, and not until World War I did the major farm groups have permanent Washington offices. The emergence of the major Washington farm lobbies in the 1920s brought more order to farmer opinion, but major components of the lobby were still divided over policy alternatives. The congressional "farm bloc," started by Republicans from Iowa in 1921, created a ready audience for the farm lobby's views and acted as a "big steering committee, to get cooperation in support of measures of common interest," in the words of Senator Arthur Capper of Kansas. Still, proposals to fix farm prices outright, to dump commodities overseas, and to guarantee cost of production, among others, divided farm state politicians, farm regions, farm lobbies, and presidential administrations, obviating a consensus on farm policy. The depth of the farm crisis by 1932, however, allowed FDR to pull the disparate forces and ideas together in support of the idea of production control through domestic allotments. With the major agricultural powers in agreement, the state could broker legislation favorable to farmers.[37]

In the postwar years the Depression-era cooperation unraveled, and it became increasingly hard for the state to broker among the interests within agriculture in addition to other political interests. As early as the mid-

1950s, "the titanic struggles between the Congress and the administration, between Cotton Belt and Corn Belt, between the Farmers Union and the Farm Bureau, between the Democrats and the Republicans fettered the farm bloc's defense of what it already had." These divisions created more organizations that thought they could help the farmer better than the older organizations. A Minnesota farmer who joined the newly created NFO stressed to his governor that "one of the NFO's main objectives is to get the 'government out of farming' somewhat. Many of the liberal gov. programs have failed and the one suffering is the farmer. How can we do anything for our country when many government controls forbid us in different areas?" Reflecting the divisions in farm group philosophies, the Minnesota commissioner of agriculture told the governor that he sided with farm program advocates such as the NFU and that "until they [NFO] develop a concrete program which accepts legislative action as a basic source of providing bargaining power, they will never be in a position to render a real service to agriculture." President James Patton of the NFU thought that "bargaining power" came through production adjustments coordinated by government programs. Patton's successor maintained that "bargaining must begin on the farms. It must begin with supply management. If it does not, it is not likely that any amount of bargaining can result in adequate prices for farmers." President Oren Lee Staley of the NFO instead argued that "anyone who has the idea that some government program is going to save farmers had better forget it and dedicate all the time he can spare to enrolling neighbors in the NFO Collective Bargaining program, the one way farmers can save themselves." But in the 1950s, again underscoring the divisions between and within farm groups, large parts of the early NFO broke off to join the NFU, believing that "anything we can accomplish in NFO can better be accomplished in NFU."[38]

A representative of the Farmer's Cooperative Marketing Association in Kansas told Congressman Bob Dole that "farm legislation in the past has failed due to the fact that it has been a political football for the past 20 years. Also I am willing to admit that farm organizations with different opinions have complicated the situation." The NFO was part of this confusion, but so were commodity-specific organizations such as the National Pork Producers Council, the National Corn Growers Association, and the U.S. Feed Grains Council, formed in 1954, 1957, and 1960, respectively. Congressman Tom Foley (D, WA) reflected the frustration of legislators ready to help farmers but unsure which farm group, or faction within a farm group, spoke for farmers. Speaking of the once powerful Farm Bureau in 1969, Foley said, "I do not know how we are to evaluate presentations that come to us from the Farm Bureau when members have so many constituents who do not agree with those presentations."[39]

Brokering a farm program was not simply a matter of coping with farmer resistance to controls and the competing agendas of the farm lobby. The republic was no longer one dominated by agricultural producers, but one dominated by consumers, mostly urban—Gilbert Fite's 1981 historical survey of farmers is subtitled "The New Minority." When the AAA coordinated the killing of pigs and destruction of cotton, many of the people who consumed such items complained. In the waning months of World War II, the influence of consumers again emerged on the issue of food prices, an influence that would bedevil farm advocates throughout the postwar period. As the trend continued, the interests of the urban consumer would become more powerful than the rural producer and the state less inclined to broker profarmer, anticonsumer laws. In 1940 Mayor Fiorello LaGuardia of New York City lined up the city's congressional votes for the AAA in exchange for relief votes from the farm bloc. But as time passed, the farm bloc would have fewer and fewer votes to offer in exchange for urban help with farm programs that increased the food costs of urbanites. In 1950, 38 percent of congressmen were from farm districts; by 1960, 12 percent were. In Minnesota's First Congressional District, for example, the proportion of farmers dropped from 33 percent to 14 percent from 1950 to 1980. Harold Cooley, chairman of the House Agriculture Committee from 1949 to 1966, saw his North Carolina constituency simply dry up: when he went to Congress in 1934 half of his constituents were engaged in farming, but by the time of his defeat in 1966 only 10 percent were. When Kennedy was elected, the farm groups were still "very important" to agricultural legislation, according to an administration official. But by the time of the Carter years, another administration official could concede that "we don't spend a lot of time" seeking the support of farm groups. The farmers' greatest political advantage, sheer numbers, evaporated, or rather, outmigrated.[40]

During the 1960 presidential campaign, James Giglio explains, Willard "Cochrane had suggested publicly that Kennedy's farm program could increase consumer food prices by 10 percent. Nixon capitalized on the political blunder by claiming that Kennedy's farm program meant a food-price hike of 25 percent. The Kennedy camp spent the remainder of the campaign explaining that a modest rise in farm prices would have a negligible effect on consumers." As *Business Week* viewed Kennedy's dilemma, it was "to win farmer or housewife: That's the problem." After he left the USDA, Cochrane admitted that "this is now an urban society, and food prices are one of the most politically sensitive items in the cost of living. Any conscious, purposeful action by government to raise farm prices and thereby raise food prices brings a flood of protests down upon the one man elected by all the people, The President. . . . A conscious policy of raising food prices is considered political suicide. . . . I learned this in the fiery furnace during the 1960 campaign, and my colleagues and I in the USDA relearned it on several occasions between 1961 and 1964." He commented in the mid-1960s that the urban voters' "Fabian retreat with regard to farm program costs . . . is about over." A consumer offensive can also be seen in the farm bargaining bill debates of the late 1960s, where consumers objected to legislation that might increase the prices of farm commodities. During the debates over the 1970 farm bill the *Minneapolis Tribune* concluded that the House of Representatives "is increasingly urban-oriented and increasingly concerned that more money is spent on declining numbers of farmers than on such things as federal aid to elementary and secondary education and the Office of Economic Opportunity's war on poverty."[41]

The more influential consumers also became more organized, crowding out the farm lobby. President Kennedy sent a consumer message to Congress; LBJ created the President's Council on Consumer Interests; and President Nixon signed the Consumer Product Safety Act in 1972. In 1966 consumers boycotted supermarket chains because of rising food prices, and in the spring of 1973 labor, urban politicians, and women's and consumer groups organized a boycott of beef. The number of stories on the three major television networks about food prices jumped from 15 in 1969 to 378 in 1973. The urban consumer became a powerful force in the political debates about inflation starting in the late 1960s, so powerful that urban-oriented legislators such as Senator Ribicoff of Connecticut would attack farmer cooperatives for increasing prices. When Secretary Butz said farmers should "plow up the fence rows" to reduce food inflation in the early 1970s, he was then attacked by the NFO for "playing to the consumer." Bill Thatcher understood that the farmers' political clout relative to consumers was slipping and attacked metropolitan newspapers and big industrialists, who had "selfish motives to get cheap food into the cities, so that they have less labor problems."[42]

The Manhattan Republican Jacob Javits, for example, started making noises about checking into the farm program in 1953. When some promoted a wheat program in the 1950s paid for by a tax on millers, it was attacked as a "bread tax" by urban politicians. Rural legislators such as McGovern tried to justify the farm program to urban counterparts by arguing that it prevented larger problems in the big cities. McGovern wrote Mayor Robert Wagner of New York that he was doing his part to "keep 'em down on the farm" instead of flooding to the cities. But even McGovern's attention to the farm problem waned. His first speech as a congressman in 1957 lashed out at Benson, but his first speech as a senator in 1963 was about Cuba policy. In 1968 he revealed his greater political interest when he tried to acquire all the Kennedy delegates at the Chicago convention and win the nomination as a "peace candidate." Those who supported his 1972 candidacy were not such pillars of the New Deal order as farmers and workers, but more like the cast of Hair, as Congressman Tip O'Neill (D, MA) said. Allen Matusow, the author of an excellent book on the farm issue in the Truman years, also reflects how politics drifted away from the farmers. His book about 1960s liberalism, *The Unraveling of America*, does not even list "agriculture" or "farmers" in the index, instead focusing on Vietnam, civil rights, the war on poverty, the counterculture, black power, and the new left.[43]

An important component of the consumer lobby was organized labor, a powerful rival to the farm lobby. In addition to the establishment of the New Deal farm program in the 1930s, the New Deal labor laws helped triple the number of organized workers, and by 1940 labor was the biggest contributor to the Democratic Party. Some labor leaders such as Walter Reuther hoped to build an electoral alliance between organized labor and farmers in the postwar years, an effort symbolized by the presidential candidacy of Henry Wallace in 1948, "a watershed in the history of the American left." But the alliance was always unlikely. Farmers paid higher prices for machinery and other supplies because of union wages, and workers paid higher taxes or higher prices, or both, for food because of the farm program.[44]

The NFU and the CIO did cooperate on the Brannan plan, labor favoring its mechanism for keeping consumer prices low. National Democratic Chairman Howard McGrath, according to Allen Matusow, saw the Brannan plan as "an opportunity to cement the tentative alliance of farmers and laborers that elected Truman in 1948." As a result, Congressman Clifford R. Hope of Kansas called it a "CIO bill from start to finish," a charge that alienated southern political leaders normally supportive of generous farm bills. Labor further jeopardized the potential coalition during the Korean War by opposing the farm parity prices of the Defense Production Act in fear of higher food prices. Farmers criticized Brannan for catering to consumers and labor in his plan and for saying that farm legislation needed to be in the "national interest, and not specific class legislation."[45]

George McGovern's early political career is again useful, as it highlights the tensions between the labor and farmer constituencies. McGovern attended the 1948 Progressive Party convention and supported Henry Wallace; as a Ph.D. candidate in history at Northwestern, McGovern wrote his dissertation on the 1914 massacre of miners in Ludlow, Colorado, by Rockefeller's hired guns. Only three months into his freshman term in Congress, McGovern went to the floor of the House of Representatives and commemorated the forty-third anniversary of the Ludlow Massacre, arguing that his reading and study of the subject in graduate school "convinced [him] of the vital relationship between a strong, healthy labor movement and effective political democracy." Knowingly or unknowingly citing the countervailing-power notion of the liberal economist John Kenneth Galbraith, who would later advise him during his presidential bid, McGovern concluded, "there is no doubt in my mind that the high degree of organization of management in our industrial society makes it imperative that working men and women be well organized if either political or economic freedom is to survive in this land."[46] McGovern wasted no time getting involved in union conventions, speechmaking, and contribution seeking. In his first month in office McGovern attended a Legislative Institute organized by the Textile Workers Union of America. Two months later McGovern participated with Chicago mayor Richard Daley in the Civil Rights Conference of the International Union of Electrical, Radio, and Machine Workers (IUE). In Sioux Falls in 1958, McGovern participated in a Farmers and Workers Conference, a meeting to promote the Brannan coalition of workers and farmers.[47]

The courting of labor was politically risky for McGovern. A leader in the Taft-Hartley movement was South Dakota congressman Francis Case, who in 1946 successfully guided a bill through the House of Representatives limiting strikes that disrupted production and distribution necessary to the "public interest," a move that one labor publication labeled "one of the most vicious attacks on the labor movement in U.S. history." The postwar strikes in the railroad industry and coal mining coupled with anger at the high-wage policies of the Office of Price Administration created much support for Case's efforts in South Dakota. During the postwar coal strike, many of Case's constituents denounced John L. Lewis as a "common thug" who pushed Congress around, and many farmers objected to high industrial wages that increased their machinery costs and other expenses. In the spirit of Taft-Hartley, one South Dakota man argued that people "should never be forced to join an organization not to our liking in these United States as long as it remains a democracy." South Dakota would be one of the first states to take advantage of the Taft-Hartley legislation and establish itself as a right-to-work state.[48] Also, in early 1957, only months after McGovern's first election to Congress with the help of union funds, Senator McClellan's "Rackets Committee" started investigating corruption in labor unions, a committee on which Senator Mundt was the ranking Republican and which therefore generated great interest in South Dakota papers. Mundt was also a member of the Senate Republican Policy Committee, which targeted the connections between labor and the Democrats in the 1956 elections as an issue to be used in the 1958 elections. The "Rackets Committee" finished work six months before the 1960 elections, and the issue would prove to be a major factor in the Mundt-McGovern Senate race.[49]

Before McGovern's 1958 congressional race was over, his support for labor unions emerged as an issue. One strategy memo argues that the most important focus of the campaign of his opponent Joe Foss would be McGovern's prolabor record. At a fund-raiser at the Palmer House in Chicago, Foss denounced union shop requirements that workers be forced to join a union and criticized the irresponsible spending of workers' union dues. McGovern responded by saying he was the first member of the South Dakota delegation to "repudiate" the corruption of labor leaders, congratulated the Rackets Committee for its efforts, supported labor reform, and pledged he would "do everything in [his] power to fight corruption in the American labor movement." McGovern in turn accused Foss of "pal[ling] around with his friends in the Pentagon and the Rockefellers," turning "up his nose at the dollars of American working men," and obscuring the "big contributions that have been flooding into his campaign from the fat cats." Invoking the image of the farmer, McGovern righteously declared, "I hope I never get so proud that I refuse political support from the man in overalls." A few weeks before the election McGovern emphatically denied an editorial in the *Pierre Capital Journal* that contended he favored repeal of the state's right-to-work law. He argued that the "Right-to-Work Law in South Dakota is not an issue in the 1958 campaign" and told the Mitchell Chamber of Commerce that he thought it was a state issue and therefore he shouldn't get involved. The labor issue never materialized for Foss, and with another strong showing among farmers, McGovern won the race by three thousand more votes than he had won in 1956, along with forty other new Democratic congressmen from rural districts that were formerly Republican.[50]

Despite the divisions between urban labor and rural farmer politicians, they could sometimes arrange mutually beneficial logrolls in Congress. In the 1950s, for example, McGovern worked to include food stamps in the USDA budget, which gave urban legislators a reason to vote for the farm bill. Urban liberals demanded that the advocates of farm bill legislation vote for the repeal of Taft-Hartley in 1965 in exchange for their support of the 1965 farm bill; in 1973 they offered farm bill votes in exchange for a minimum-wage vote. But the arrangement was always tenuous. McGovern voted no on the repeal of Taft-Hartley, for example, and he clashed with labor over the issue of grain exports to the Soviet Union in the 1960s. To gain labor support for the exports, the Kennedy administration had to promise that 50 percent of the shipments would be made on American ships. McGovern told Ralph Massey of the International Longshoremen's Association that the requirement cost farmers $250 million in wheat sales in 1965. When McGovern ran for president in 1972 these legislative battles hurt him—organized labor was suspicious of him because they suspected he had "tacitly supported right-to-work laws."[51]

Labor's primary objection to the farm program was that it increased the costs of food to workers, a criticism that diminished with the Brannan plan and the post-1964 farm bills, which allowed prices to fluctuate and made payments to farmers to boost their income. But many farmers objected to the payment system. In 1948 the Farm Bureau opposed compensatory payments as welfarelike and subject to the budget whims of Congress. The NFO later agreed, opposing the 1970 Nixon legislation because "it makes farm income more dependent upon government payments, and then leaves this income more vulnerable to proposed legislation to limit payments."[52]

In addition, this system was expensive, pitting farmers against other interests pursuing funds from the federal government. If the Brannan plan was really to keep retail food prices low and farm income high, it would take large government expenditures—up to $8 billion a year, some predicted—a level of spending difficult to sustain. When McGovern went to Congress in 1956 the USDA spent $1.6 billion on farm programs; by 1959 it spent $5.8 billion, behind only the defense budget and debt servicing, an amount severely criticized. One of the key rationales behind the Kennedy administration's production controls proposal was to bring down the cost of the farm program while maintaining farm income. But with the adoption of the voluntary programs, costs increased to close to $8 billion a year. Willard Cochrane acknowledged that "everyone would be happy with voluntary control programs if it weren't for one little item—program costs." Payments needed to be large enough to entice acres out of production, and with growing agricultural productivity, payments increased because the number of acres that needed to be pulled out of production grew. In 1969, with the success of the farm program linked to its share of the national budget, John Melcher of Montana won election to Congress in Montana by accusing Agricultural Secretary Clifford Hardin of "representing the Budget Bureau and not the farmers." And 1969 was also the last year the country balanced its budget, so pressures to cut government spending, to balance the books, and to ease inflation, coupled with the 1970s "tax revolt," put additional pressure on the farm program. James Bonnen said the state was faced with a choice of high budget costs or passing the costs on to consumers through strict production control programs, which both consumers and farmers didn't like.[53]

Other issues and interests also had to be considered. The farm program's effect on international agricultural trade became an issue when it was feared that the program cost the nation foreign sales. The master of the National Grange, for example, argued that farmers suffered from what he called "market shrinkage," or the sales that could have been made if price supports were lower. Public support for production controls was also thin during the period after World War II, when many feared the possibility global starvation, or during the 1960s, when it seemed that millions of people might starve in places such as India and Biafra. As a Farmers Union official pointed out in the 1960s, "the world's last great famines occurred twenty-two years ago, in Bengal in India, and in Honan Province in China. Millions of people died of starvation. . . . Up to the 1940s famines of megadeath proportions—with people dying by the millions—were quite commonplace in the world. . . . When famine threatened many times since then in various parts of Asia, Africa, and South America, our stocks of wheat have come to the rescue and saved millions and millions of lives," a fact that hindered efforts to cut food stocks. The politics of the farm program, then, often drifted into international relief questions, foreign policy matters, and therefore cold war rivalries.[54]

The agricultural processing industry was also affected by farm policy but often received little political hearing. John Hansen has studied the "expulsion of the farm processor interests in the late 1920s" and how "Congress treated the agricultural trades with antagonism and scorn." In 1933 the processors were taxed to pay for the AAA because they benefited from the "farmer's inability to check his operations" and his disorganized production and marketing practices. In 1947, at the House Agriculture Committee's "regular hearings on long-range price-support policy, it treated the handful of tradesmen who dared show up to a morning of bipartisan business-bashing." The weakness can also be seen in the failed lobbying efforts of the National Tax Equality Association. When the farmers and farm groups were deeply divided during the farm bargaining bill debates, however, the processors had greater clout.[55]

By the mid-1960s it was possible to establish an equilibrium between farmers, laborers, and consumers, but one that was made possible by the tax revenues of the state, and therefore one vulnerable to budget cutbacks. The farm recession of the 1980s placed enormous pressure on the state to provide money to cover the gap between farm prices and government target prices, a situation more precarious for farmers owing to worsening federal budget deficits starting in the 1970s and wider public support for tax cuts. The situation rekindled the idea of Kennedy-like mandatory supply control programs such as the Harkin-Gephardt Save the Family Farm Act, but such measures failed to gain enough support to pass. In 1990, after the worst of the farm recession had passed and public attention diminished, several political leaders made efforts to dismantle the farm program and succeeded in terminating the honey and wool programs. In 1996 the "Freedom to Farm Act" was enacted into law, phasing out the corn, soybean, and wheat programs over the next several years.

During most of the postwar period, the farm program only partially succeeded in organizing farm marketing. Wheat farmers voted for moderate production quotas in the 1950s, for example, but resisted the tighter controls needed to increase prices through artificial scarcity, undermining the "acceptance or legitimacy of public interest and private restraint" needed for political organization. Farmers' failure to achieve everything they wanted partly reflected their disagreement over which legislation was beneficial and partly reflected ideological opposition to state control of their economic lives. Coupled with the opposition of interest groups such as labor and consumers and the declining political influence of farmers, such sentiments complicated the state's efforts to fully organize farmers. Many farmers preferred to organize their marketing activities themselves, or through the NFO, other farm organizations, or farmer cooperatives, seeking the best prices they could find.

CONCLUSION

American farmers' critique of the monopoly problem persisted into the last half of the twentieth century. Many farmers feared that monopolistic corporations manipulated farm prices and would ultimately take over agricultural production, a reasonable fear considering the upheavals in postwar rural life. But the takeover did not materialize, the agricultural processing sector displayed some signs of horizontal competitiveness, and when farmers sought to organize their marketing and improve their bargaining position they were sometimes successful.

The investigations of corporate farming operations indicated that a small percentage of grain belt farming was corporate (less than 4 percent, according to the 1992 Census of Agriculture), and what was corporate was often owned by family farmers.[1] But the fears generated by the prospect of corporate farming underscore the noneconomic and decentralization dimensions of antitrust law. Many farmers and rural advocates called for antitrust action because they regretted the changing scale of agriculture, which reduced the number of farmers and therefore reduced the number of small towns that survived as service centers for farmers. Instead of living on farms and in small towns, people moved to the cities in grain belt states or to metropolitan centers around the country, ending the rural communities and traditions so many Americans had known and cherished.

The food processing sector was blamed for many of these changes because of its "monopolistic control" of farm markets. But in the formal economic sense, this sector displayed some competitive characteristics in the postwar years. New firms entered the meatpacking sector, for example, and pushed many older firms toward bankruptcy. The meatpacking sector's potential for monopolistic behavior was also constrained by the feeding and retailing sectors, unstable demand, and competition from other products. Wheat milling and soybean processing also displayed signs of competitiveness and were highly dependent upon the wide swings in international prices. Firms that exported farm commodities faced barriers to collusion and competition from farmers who decided to export through the NFO, cooperatives, and other organizations.

Processing firms also competed over the acquisition of farmers' products—and farmers during this period engaged the processing sector in a more organized fashion. The NFO constantly emphasized the importance of increasing farmer bargaining power and pooling commodities for collective sale to processors. One farmer remembered that "the NFO taught me to bargain."[2] Farm marketing cooperatives such as Farmland, GTA, and Far-Mar-Co grew substantially and even began to process and export farm commodities. Farmers also organized their marketing through the state, often voting to impose production controls on themselves in order to receive a higher price for their products.

One of the core organizational problems for farmers, however—one not taken very seriously in historical political economy—was the libertarian streak of many farmers. Some farmers resisted the controls associated with the federal farm program from the beginning, actually voting down the program when the Kennedy administration tried to impose strict production controls. One farmer told George McGovern, one of the greatest promoters of the farm program in Congress, that he'd rather "take [his] chances with jungle economics" than with the government.[3] A Kansas farmer complained that he "spent no less than an accumulated 15 days in 1961, running to and from the County ASC [Agriculture Stabilization and Conservation] office, getting information, signing papers, making appeals on allotments & yields & trying to make decisions. It seems that our present administration is hungry for controls & more controls."[4] Many farmers also rejected the controls associated with membership in organizations such as the NFO and the farmer cooperatives, even though the organizations argued that they were trying to improve the business skills of farmers and more effectively market farmers' products. The NFO, in many ways the most radical of the postwar farmer movements, underscores the libertarian tendency. NFO members were not radical socialists or statists—they harshly criticized the federal farm program—but were capitalists, promoting business acumen and better marketing strategies among farmers and opposing mandatory bargaining regimes. As one newspaper said, the NFO was trying to "fight capitalism with more capitalism."[5]

When farmers did not accept greater economic controls, USDA officials such as Willard Cochrane believed it was because "farmers generally [did] not understand or comprehend the nature of the farm program of which they are a part," evidence of the "distressingly low" "economic literacy of farmers." He saw farmers as "hopelessly divided and badly confused before their problem; their political power is dissipated and their understanding is muddied."[6] Cochrane, the same man who thought agriculture should be regulated as a "public utility," saw farmer opposition to controls as simply a matter of confusion and muddy thinking, even though a *Farm Journal* poll a few years earlier indicated that 55 percent of the nation's farmers opposed government involvement in agriculture.[7]

Such libertarian attitudes can be seen as another component of the monopoly problem—resistance to outside economic control, public or private—but not as a willingness to ignore the moral implications of economic change in rural America. The farmers protesting the monopoly problem in the postwar years feared the social costs as much as the economic. In the debates about corporate farming, for example, some arguments were made about the efficiency of the family farm, but the broader argument was that corporate farms destroyed rural communities and displaced proud, honest, hardworking farmers, adding to the weakening effects of growing urban consumerism in postwar America.

After so many farmers left the land, we became a nation of "mollycoddles," as Ben Hogan would say. The poet and cattle producer Linda Hasselstrom, to take another example, sees the social costs manifest in her neighbor: "Like many ranchers of forty or older, Bill was raised to do everything as well as possible, and believes any work done with pride is enjoyable. I consider his kind a vanishing species in an America that used to teem with pride in all its labors, and found enjoyment in that pride, and that work." She says that "people whose lives are a challenge are healthier in every way; by taking the difficulties, the tests, out of life, we've turned it into oatmeal."[8] Another man who frequented the northern parts of the grain belt would agree. One of Teddy Roosevelt's most famous speeches was about "the strenuous life," about "those virile qualities necessary to win in the stern strife of life." As Christopher Lasch has written about the demoralizing and deadening effects of modern American liberalism and consumerism, "if humanity thrives on peace and prosperity, it also needs an occasional taste of battle." Hasselstrom wonders about what physicians call "'diseases of civilization,' consequences of our less active lives."

> We take a body with [a] history [of physical survival], prop it upright
> for eight hours while the fingers lightly punch buttons, then seat it in a

car where moderate foot pressure and a few arm movements take it home. It hugs the other members of its small tribe, then slumps down on a cushiony surface and aims its eyes at a lighted screen for two to six hours, and lies down on another soft surface until it's time to get up and do it all again. No wonder we're sick.[9]

Hasselstrom pities city folks forced to find artificial means of exercise. "Armed with the strengthening effects of fresh air and exercise, and saved from the mental problems created by urban stress and overcrowding, country people who regularly do physical labor are healthier and saner than anyone in the city, and we'll outlive and outsmart our critics." Instead of shrinking, the cattle producers Hasselstrom knows master their adversity. Her husband "could accept whatever came his way later in life; perhaps it had something to do with his patience, his calm, his determination to make the best of his life, and of his death."[10] For Hasselstrom, these qualities were on display every morning in the "Coffee Cup Cafe":

Soon as the morning chores are done,
cows milked, pigs fed, kids packed
off to school, it's down to the cafe
for more coffee and some soothing
conversation . . .

So for an hour they cheer each other, each story
worse than the last, each face longer. You'd think
they'd throw themselves under their tractors
when they leave, but they're bouncy as a new calf,
caps tilted fiercely into the sun.

They feel better, now they know
somebody's having a harder time
and that men like them
can take it.[11]

Hasselstrom, in what could be a classical republican critique of luxury's corrosive effect on virtue, regrets the attitudes of the children of some cattle producers, consumed by "their desire for instant and immense gratification in the form of electronic gadgets, a bigger car, [and] a designer home [which] has forced them to live in towns where they can earn more." Like Christopher Lasch, Hasselstrom sees virtue in the petty bourgeoisie and their sense of limits, recounting that "older ranchers were raised with the philosophy that people should not expect to get more than they can make or grow with their own hands and sweat, an idea scorned in our consumer society." In contrast to the rancher philosophy, Lasch fears the growth of an elite social class in the United States, what Robert Reich calls information-age "symbolic analysts," which "maintain[s] the fiction that its power rests on intelligence alone. Hence it has little sense of ancestral gratitude or of an obligation to live up to responsibilities inherited from the past. It thinks of itself as a self-made elite owing its privileges exclusively to its own efforts."[12]

To avoid such an existence and such attitudes is to avoid urban America. But fewer farms meant that more farmers migrated to the big city, which, as Garry Wills notes, "in the American imagination has played roughly the role of hell in Christian theology" (partially explaining why, in one postwar survey, two-thirds of farmers thought forcing farmers to leave their farms was "un-American and even non-Christian").[13] Many of the postwar letters of farmers and farm wives uncertain of their economic future feared a potential city life, echoing the poet of the 1880s:

The city has many attractions,
But think of the vices and sins,
When once in the vortex of fashion,
How soon the course downward begins.[14]

The city is burdened, as another verse goes, with

Its cries of want and wild despairs;
Its dust and smoke which stifle breath;
Its foul effluvia of death;
Its catacombs of human lairs[15]

The rapid depopulation of rural America roughly corresponded with the beginning of the contemporary American culture wars, after which many wondered if the republic could survive the erosion of the older attitudes, values, and traditions. People questioned, as Spiro Agnew famously said, the tendency to "sneer at honesty, thrift, hard work, prudence, common decency, and self-denial" in favor of the "permissiveness that in turn has resulted in a shockingly warped sense of values." This contrasted to, say, the disappearing "Nebraskan character, founded upon the unglamorous virtues of common sense, reticence, compression, and reserve"; or in another version, Nebraskans as "plain, sensible, honest men, who have never begged any odds in the game of life, and whose strongest wish seems to be to stand square with their fellows." An Iowa farmer noted the contrast when describing farm protests in a letter to Congressman John Culver (D, IA), asking him "how many long-bearded militant farmers have stormed your office in Washington asking for the impossible? How many draftcard-burning and flag-burning farmers have picketed the Ag. building?"[16]

These social, moral, and character questions, tightly bound up in the monopoly issue, tend to be slighted by intellectuals. Strains of European culture and thought have always scorned the farmer:

> Stolid and stunned a brother to the ox
> Whose breath blew out the light within that brain?
>   What to him
> Are Plato and the swing of Pleiades?
> What the long reaches of the peaks of song,
> The rift of dawn, the reddening of the rose?[17]

But even in the United States, the haven of the yeoman farmer, the literati began to sneer in the late nineteenth century. As C. Elizabeth Raymond has written, rural life was increasingly seen as "smug, socially backward, and intellectually confining." Mild joking about farmers turned "savage," according to David Danbom, and "such derogatory labels as hick, rube, and yokel became regular parts of public discourse about people who were defined as distinctly—and perhaps dangerously—inferior." In 1903 an Eastern writer thought of the typical farmer as "a lean, gawky, bewhiskered creature, ignorant of all topics that lie outside the sphere of farms and crops." Presaging present-day coastie snobbery, Edmund Wilson, fresh out of Princeton, complained about traveling to California from the East by saying that the "trouble is that you have to pass through the Middle West on the way, and I wouldn't be sure of the felicity of any union under those auspices. The children of such a union would be morose and deformed." Tired of the moralizing farmers injected into American politics, H. L. Mencken said, "We'll all be better off when the men who raise wheat and hogs punch timeclocks," agreeing with Marx that the "enormous cities . . . [have] rescued a considerable part of the population from the idiocy of rural life."[18]

The emergence of the counterculture, its currency in the academy, and the hostility to "bourgeois virtue" divert attention from the benefits of rural life. The neglect and "anti-rural bias" of the "condescenders" in the historical profession partially explains why we know so little about what happened to farmers after World War II and why the midwestern historical "landscape is quite barren."[19] A University of Iowa English professor complained in the 1940s that

> the "American Mind" takes up its position on the rim of the country and looks in. Even to many intellectuals physically resident in Iowa, Iowa seems more "remote" than Connecticut or Brooklyn. Often, in journalistic and political discussions, New York seems to block out everything farther west; in academic discussions, even now, Massachusetts is still occasionally allowed to perform that unhappy role. In no other country is the thinking so likely to be peripheral.[20]

The neglect also stems from the fact that grain belt farmers living the "strenuous life" do not fit the model of human behavior many scholars favor. Currently, an academic project attempts to understand why people haven't revolted and attacked the concentrations of economic wealth in capitalist democracies, what can be called the monopoly problem, the evils alluded to by Ben Hogan and other farmers. Work in this area follows the theories of Antonio Gramsci, who attempts to "explain why workers under advanced capitalism have not behaved the way Marx said they would." The powerful in society, so the story goes, so dominated the culture, ideas, and discourse that revolutionary ideas and therefore revolution were subverted. Historians set to work in recent decades trying to explain how the hegemonic system functions and to identify dissidents. Some people have resisted the cultural hegemony of liberal capitalism, so some historians claim to have discovered, finding, as John Patrick Diggins derisively notes, what "Tocqueville and Orestes A. Brownson, Emerson and Thoreau, James Fenimore Cooper and Herman Melville, Veblen and Henry Adams, Charles A. Beard and Walter Lippman, and almost all other American intellectuals, including expatriates like Santayana and even émigrés like Hannah Arendt and Herbert Marcuse, so consistently failed to find in America."[21]

The story of postwar American farmers causes one to wonder about the recent claims of these new historians. Farmers seem to substantiate older intellectual interpretations, given their hard work, striving, enterprise—Hogan working his kids hard, keeping his fences "horse high, bull strong and hog tight," souping up his tractors so "he could plow or disc or cultivate more acres per day than anyone else in the county, avoiding too much borrowing, inventing new machinery, balancing a diverse portfolio of milk cows, corn, soybeans, sheep and turkeys," and searching for the best price for his product, like the NFO and its members. Above all, Hogan worked hard: "He worked and never slowed. He bulled his way through the house before sunrise each morning, growling to his sons to get out of bed and do the chores." It is the kind of attitude that John Miller sees embodied in farmers like Pa Ingalls, representing the "typical frontiersman—willful, self-sufficient, industrious, and above all individualistic." In the documentary film *Troublesome Creek*, an older Iowa farm couple decides to sell out because they think the bank may soon call their loans. They had worked the farm for sixty years, and worked it well, and the sellout seemed unjust. People in the film hated the banker, the icon of the capitalist class, but they did not man the barricades and, despite their dislike, made it a point to be nice to him. They accepted and adapted, similar to the way Linda Hasselstrom described her husband's grace. The implication is that maybe America is the way Louis Hartz, writing forty years ago, described it. Social theories that privilege status, elitism, power, class, race, or gender as categories of analysis, methods of explanation, and theories of causation overlook what Hartz called the "liberal tradition," or what Richard Hofstadter saw as the enduring political framework encompassing the "belief in the rights of property, the philosophy of economic individualism, the value of competition."[22]

Americans chose as their "favorite star" in 1995 John Wayne, whom Garry Wills calls the "most recent embodiment of that American Adam—untrammeled, unspoiled, free to roam, breathing a larger air than the cramped men behind desks, the pygmy clerks and technicians. He is the avatar of the hero in that genre that best combines all these mythic ideas about American exceptionalism—contact with nature, distrust of government, dignity achieved by performance, skepticism toward the claims of experts." Understanding Hasselstrom and Hogan and other farmers may help explain, as Wills says, why "modern intellectuals are puzzled by the popularity of John Wayne."[23] Were we really a bunch of fools blind to our own oppression by capitalist cultural hegemony, or were we a free people, working to improve our lives, celebrating the successes of a republican form of government that had never worked before, and pursuing happiness? Maybe we should admit that the Iowan John Wayne can tell us more about our country than the Italian communist Antonio Gramsci.

It would be a mistake to overlook the sense of loss felt by many farmers and to only interpret their experience as one of co-optation by the capitalist hegemony. It is an increasingly common oversight in historical interpretation, however, as the case of the Puritans makes clear. David Harlan recently explained how the great historian Perry Miller bequeathed a body of knowledge about the Puritans that explained their hopes, exposing readers to the complexities of the "New England mind" and its struggles with its own failings. As Harlan explains, "Almost single-handedly, Miller transformed seventeenth-century Puritanism from a dull and barren period into one of the most vital and important episodes in the history of American thought, a source of rich insight into human nature and penetrating criticism of American culture." Miller's successor at Harvard, Sacvan Bercovitch, can see only the elements of social control and cultural hegemony in Puritanism, however. Harlan explains that "Bercovitch approached American Puritanism not as a source of insight but as a system of deception. . . . Puritan books seem to him little more than empty ciphers, incapable of exposing moral illusions or providing moral guidance. . . . [Whereas] Miller saw something of deep and abiding value in the Puritan tradition: an acceptance of melancholy and sadness as ways of seeing, almost as signs of grace."[24] Harlan's warning warrants attention lest historians interpret the changes in farming as mindless acquiescence to economic change and fail to detect the sense of loss and pessimism among farmers leaving the land.

Farmers weren't protesting the system or praying for revolution so much as they were hoping their farms, small towns, and rural communities could survive the turbulence of economic change. One woman who regretted the changes in rural life and "saw [her] parents lose their farm during times which were truly difficult" proudly remembered that despite the pain involved, "at no time did [her] parents resort to lawlessness and violence."[25] Most farmers were not willing to support radical alternatives—socialism, say, a tendency dating back to Populist protests over the monopoly problem. As LeRoy Asby has written, for example:

> Socialism held little more attraction for many Populists than it did for Bryan. They prided themselves as landowners or aspiring landowners, and they fretted that society would falter without the incentives of profits and private property. According to one leader of the third-party movement, Tom Watson of Georgia, Populists rejected "Socialism, with all its collective ownership of land, homes, and pocketbooks."[26]

Farmers' capitalistic tendencies can be seen since the beginning of the republic, and in the activities of their ancestors in fourteenth-century England for that matter, part of a persistent "capitalist hunger" in America, according to Hartz, "despite talk of 'monopoly.'"[27]

Farmers in the postwar years were trapped by what Daniel Bell calls the "cultural contradictions of capitalism," or the conflict between social realms, the "techno-economic," the "polity," and the "culture." Farmers were largely individualistic, opposed to onerous statist controls, and capitalistic, conferring legitimacy on economic change. But their regrets about such changes can be seen in calls for social justice in the polity realm and in their sorrow over the cultural costs of a dwindling number of farmers and small towns, outcroppings of the American republican tradition, too often overshadowed by the more powerful liberal tradition. In this "disjunction of realms" Bell sees "many of the latent social conflicts that have been expressed ideologically as alienation, depersonalization, the attack on authority," and, it might be said, the monopoly problem. The antimonopoly tradition can be seen, according to the economic historian Stuart Bruchey, as "a cry of protest on the part of individuals increasingly depersonalized and lost in corporate anonymity, of small towns increasingly invaded by the railroad, of small business and small farmers increasingly menaced by large-scale and distant competition."[28]

The value and moral questions inherent in the monopoly problem underscore the need to consider the noneconomic aspects of antitrust such as farmers' organizational efforts. While a different antitrust policy will not necessarily increase the number of farmers or revive many small rural towns, it may promote the relative organizational strength of farmers. Although sometimes difficult to justify on competitive grounds, a tighter merger policy, for example, may aid farm income by maintaining a larger number of firms for farmers to bargain with. Such a policy would avoid the statist controls many farmers despise, more fully recognize the political intent and uses of antitrust, and advance the goal of decentralization Publius promoted as a core component of a stable and functioning republic.

Using agricultural history to recall the prominence of the monopoly problem in American political history also helps contextualize many contemporary political questions, ones too often removed from their historical roots. The issue of corporate power became a media event in early 1996, for example, as large-scale corporate downsizings generated headlines, as did Patrick Buchanan's successful attacks on corporate greed in the early Republican presidential primaries. The issues of campaign finance reform and the influence of tobacco firms fed the discussion, along with federal antitrust action against Microsoft and the payment of a record antitrust fine by Archer Daniels Midland. The unprecedented corporate mergers and rapid technological change of the 1990s has generated concerns even in business circles. A *Business Week* editorial announced that "it is time to start worrying about monopolies again," and the *Wall Street Journal* ran a long series on mergers entitled "Amalgamated America." The Rockefeller biographer Ron Chernow has noted that the "swelling tide of consolidation has provoked a militant mood in Washington," one "reminiscent of the early 1900s."[29]

Such pressures contribute to a continued debate about the proper role of the antitrust laws and, in particular, continued doubts about the legitimacy of economic efficiency as the basis of antitrust doctrine. Peter Carstensen, for example, concludes that the economic efficiency arguments against antitrust activity "lack an empirical basis," explaining that the attacks focus on economic theory and legal doctrine but then proceed to make large-scale conclusions about economic history. But, as Carstensen notes, "Mere proof of the illogic of doctrines or decisions does not demonstrate what historical consequences they may have had." He uses many different examples, including ones from the oil, meatpacking, banking, and steel sectors, to indicate that antitrust activity that stopped mergers may have ultimately produced a more competitive economic sector.[30]

While emphasizing the abstractness of the arguments marshaled by the anti-antitrust commentators, Carstensen also concedes that his model of "historical evaluation leads to weak conclusions," concluding that his "studies also show that the claim for any significant positive effect on performance may be as tenuous as critics' claims of major efficiency costs," especially given recent doubts about the economic benefits of many corporate mergers. The inability to reach any solid "scientific" conclusion about the economic effect of the antitrust laws corresponds with some recent work underscoring the limits of economic expertise. And without reasonably reliable expert opinion, the monopoly problem becomes largely political, left to the normative conclusions of legislators, a proper conclusion given the profound political questions embedded in the Sherman Act.[31]

Rudolph Peritz follows in this line of criticism, arguing in favor of "escap[ing] from the prisonhouse of microeconomics" and urging "policy makers to recognize the impact of economic power on citizens, consumers, suppliers, rivals, and sovereigns." He offers a persuasive rationale for making "bigness" a more important factor in microeconomic analysis, warning that "the steady din of competition rhetoric has numbed our faculties and kept us from remembering that competition policy has never been the sole normative ground for antitrust laws." Some free market advocates even admit that "large numbers of working people and their intellectual surrogates still feel in their bones that an unfettered free market is a jungle, that workers do not get their fair share of what they produce . . . [and] that it leaves undone or poorly done all the things a good society needs most." The foreboding was recently evidenced by a South Dakota hog farmer: "You look at the Chrysler merger. You see Citibank and Travelers getting together. What's going to happen when all this money is concentrated in so few hands?"[32]

Another core ingredient of a successful republic mentioned by the American founders, one most often identified with the yeoman, cannot be helped directly by antitrust policy. "Measures designed to assure the broadest distribution of economic and political responsibility," such as antitrust, help democracy in certain ways, but more important is the maintenance of some form of civic and personal virtue. "Formally democratic institutions do not guarantee a workable social order," as Lasch argues, hoping for more attention to the "hitherto neglected traditions of thought, deriving from classical republicanism and early Protestant theology, that never had any illusions about the unimportance of civic virtue" and to thinkers "who understood that democracy has to stand for something more than enlightened self-interest, 'openness,' and toleration." Effective democracy, he believes, "requires us to speak of impersonal virtues like fortitude, workmanship, moral courage, honesty, and respect for adversaries" so we can hold each other accountable, because "unless we are prepared to make demands on one another, we can enjoy only the most rudimentary kind of common life." Without this ingredient, according to Lasch, democracy becomes unworkable, suggesting "the need for a revisionist interpretation of American history, one that stresses the degree to which liberal democracy has lived off the borrowed capital of moral and religious traditions antedating the rise of liberalism." Focusing on this dimension of American history addresses David Harlan's complaint that "there is no sense of urgency in American historical writing, no sense that we must use the books and ideas we have inherited from the past to put our own lives to the test."[33]

The political journey of George McGovern, originating with the farm problem, illustrates the drift away from the social traditions that Lasch values. When McGovern was elected to Congress as a defender of the New Deal farm program in the 1950s, one farm family living near Chancellor, South Dakota, in a county settled by politically and socially conservative Germans, Danes, and Russians, wrote a letter to McGovern about the farm problem and at the bottom scribbled, "We like to have religion in school," presaging the Supreme Court's controversial school prayer decision a few years hence. As a Methodist minister's son who had attended the seminary and used Christian teaching in his speeches, McGovern hardly seemed a threat to the farm family's views. When he was growing up in the 1920s in his hometown of Mitchell, kids went to the cowboy movies at the Lyric Theater, listened to country songs on WNAX, played marbles, bummed rides with local farmers out to the "Jim" River to go fishing, and played on the beach at Lake Mitchell. The local Corn Palace hosted John Philip Sousa playing "Stars and Stripes Forever," and during Corn Palace Week people took carnival rides and watched Sonny Boy Campbell dive off a fifty-foot platform into a tank of water three feet deep. In 1964 the former Mitchell boy turned U.S. senator confessed his continuing squareness to a high school graduation class: "I, for one, don't quite dig the Beatles."[34]

But from about the time of that speech a series of cultural divides opened up in the country, and by 1980 McGovern seemed to be on the wrong side of all of them. Suddenly McGovern was less identified with the plight of the farmer and small towns than he was with—in the words of his old friend, liberal stalwart, and farmer champion Hubert Humphrey (also born and raised in South Dakota)—the three As, "Acid, Abortion, Amnesty." Instead of the 1956 campaign based on bashing Ezra Taft Benson and pleading the case of the farmers with funds and help from the Farmers Union and a few CIO locals, McGovern's 1980 campaign was financed by Hugh Hefner, Tom Hayden, John Denver, Paul Newman, Burt Lancaster, and Marlo Thomas. Reagan won South Dakota in 1980 with a 63 percent plurality, and McGovern lost his Senate seat.[35] The social and cultural changes in America that took place over the span of McGovern's career, especially the continuing slide into consumerism and narcissism, coupled with the depopulation of farmers and the failure of many small towns, contributed to the nation's "vague sense of loss," even among urbanites. According to James Shortridge,

> Instead of generating wholesale condemnation by writers, small towns and traditional farms, indeed the entire Middle-western culture, began to be labeled quaint. Support for this viewpoint quickened in the mid 1960s, and by the early 1970s it was perhaps the dominant image that outsiders held about the region. From this perspective, the Middle West had become a museum of sorts. No up-and-coming citizen wanted to live there, but it had importance as a repository for traditional values.... A society increasingly complex, mobile, and ava-

ricious was beginning to yearn occasionally for simplicity, virtue, and rootedness.[36]

McGovern's early political focus on farmers and preserving farm life was wise, and nostalgia was justified. An academic study by the agricultural economist Luther Tweeten concludes that "compared to the general population, the farm family is more stable and the typical farmer more religious, politically more conservative, and happier and more satisfied with some aspects of life" and that "farmers are among the better-adjusted members of society. They are optimistic and have a healthy outlook on life both in terms of interpersonal relationships and general viewpoint."[37] Garrison Keillor explains it more poetically:

> What truly distinguishes Minnesota isn't majorness or hipness but a sweetness of character. . . . This is a state of people not so far removed from the farm, and farming is a civil business that believes in sharing new information and helping your neighbor. It produces goodhearted people who are tolerant, helpful and friendly. Farming is why the narcissism quotient is low here, and people avoid stupidity when possible, not wanting to be a $10 haircut on a $.50 head. The sort of arrogance that amuses New Yorkers is here considered gauche.[38]

Richard Critchfield doubts "whether America, having come so far from its rural roots, is governable." It is, but not without attention to the virtues Lasch mentions, many of which stemmed from the country's farming and small-town heritage. Jean Bethke Elshstain warns about forgetting such traditions: "Culture changes through the ongoing engagement between tradition and transformation. If we lose tradition, there will be no transformation. Only the abyss."[39] While the economic dimensions of the monopoly problem may have been exaggerated at times, the values inherent in antimonopoly protests also deserve attention, partially because they may legitimize another way of thinking about antitrust, one that also helps preserve the integrity of the republic.

EPILOGUE: TOWARD AN AGRARIAN ANTITRUST

Farmers continue to place great hopes in the antitrust laws. Throughout the 1990s farmers have sought greater antitrust enforcement as a method of alleviating the abuses of large buyers of agricultural goods, drawing on recent evidence of concentration to make their case for antitrust relief. During congressional testimony in January 1999 farmer advocates presented the results of a recent compilation of concentration data. The study indicated, for example, that five firms conducted over 80 percent of beef packing and that six firms conducted 75 percent of pork packing. Also, the four largest grain buyers controlled nearly 40 percent of the elevator facilities. Cargill, indicating the multiple product markets occupied by many large food firms, was among the dominant firms in all three markets. Congressional concern with such concentration levels, highlighted by the pending merger of Cargill and the large trader Continental Grain—termed the "mother of all mergers" by one farm group—has prompted calls for a moratorium on further mergers and acquisitions among large food firms. More generally, congressional leaders have called on the Department of Justice to "aggressively investigate concentration in agriculture."[1]

Some recent economic studies indicate a strong correlation between concentrated food firms and their profitability and market power. Compounding such concerns are widening gaps between retail and farm prices. From 1984 to 1998 consumer food prices increased 3 percent while the prices paid to farmers for the products plunged 36 percent. The impact of the price disparity is reinforced by reports of record profits among agribusiness firms at the same time that agricultural producers are suffering through a severe economic depression. This contrast in economic health between vertically related sectors, to many observers, indicates the existence of market power in the concentrated processing sector and the powerlessness of farmers.² Unfortunately for farmers, the antitrust laws have never been able to adequately address such concerns, especially the existence of a bargaining power disparity between individual farmers and large-scale corporate buyers.

Antitrust commentary deals almost exclusively with the power of sellers and injuries to consumers. Despite limited commentary, however, some courts have recently given greater recognition to potential abuses by powerful buyers. In a case involving the merger of large rice-milling facilities, a federal court did recognize that California growers would face problems finding alternative buyers. The court also noted that rice millers understood their monopsonistic position, knowing they were the "only good outlet for the California growers." Large investments in rice-growing operations limited the ability of growers to switch to other crops, conferring a greater degree of bargaining power on millers. Entry into the milling market was also deemed to be unlikely owing to the expense of building a mill and the difficulty of establishing a grower base from which to buy rice. The resulting reduction in competition for the purchase of rice stemming from the merger of major mills caused the court to find a violation of section 7 of the Clayton Act.³

Judicial recognition of bargaining power issues has also come in the form of a defense to challenged mergers. Courts have entertained the argument that a larger, more powerful firm resulting from a merger may be acceptable if the firms it sells to also possess market power. In *United States v. Country Lake Foods Inc.*, a case involving the merger of two firms in the fluid milk processing industry, the court recognized the ability of large food corporations who bought milk to check the power of milk processors. The court noted the "extremely concentrated" nature of the food-processing industry in the relevant market, where the top-three concentration ratio was over 90 percent. The size of the food firms and the volume of their purchases allowed them to monitor milk prices, making them "very sophisticated buyers." The court noted their ability to switch to other milk processors and to enter the processing market themselves. The market entry of the large food processors would be aided by their capital resources, which would allow them to purchase an existing plant, and by their existing customer base. The court found the power-buyer defense the "most persuasive argument" advanced by the defendants.⁴

Implicit in the recognition of the power-buyer defense is the assumption that powerful firms in a market can exploit small and disorganized firms in a vertically adjacent market. In other words, the power-buyer argument provides a rationale for halting the growth of powerful agribusiness processors at the expense of the thousands of farmers who sell to them. In *United States v. United Tote Inc.*, the court rejected the power-buyer defense because it recognized the relative disorganization of the buyers of the totalisator. Because so many buyers were present in the market and the buyers possessed different levels of sophistication, they could not constitute a legitimate check on the power of the sellers. In the recent case *FTC v. Cardinal Health Inc.*, the District of Columbia Court of Appeals considered the potential power of firms who bought drugs from the four largest wholesale distributors of drugs in the nation. While the court noted the power of certain buyers in the market, it also considered the numerous independent pharmacies that lacked the power to bargain effectively with the large wholesalers. The existence of a large number of buyers and the presence of many small independents created a "fragmented" buying sector unable to counter the power of the wholesalers.[5]

In tandem with judicial recognition of the importance of monopsony power, the power-buyer defense creates a rationale for scrutinizing the power of buyers relative to sellers. Thousands of farmers, for example, are often hard-pressed to muster the market power necessary to check the powerful food companies who buy their products, as the stumblings of the NFO indicate. Farmer marketing is characteristically disorganized and "fragmented," similar to the descriptions of the totalisator and wholesale drug buyers described in the *United Tote* and *Cardinal Health* cases. Because farm prices are publicly reported, buyers are also aware of any efforts to seek higher-than-market prices and can immediately switch to a different seller, dramatically lessening the chances of seller-power existing. By considering the nature of farmer marketing in agribusiness merger cases, courts could more faithfully carry out the intentions of lawmakers to promote the bargaining power of farmers.

In recent years courts have also considered the importance of information disparities in markets. Rejecting the utopian assumption of "perfect information" prevalent in economic theory increases the possibility of a more sophisticated economic analysis that takes into consideration the limited information available to individual farmers relative to the buyers of their products. The leading case in this area, which "revolutionized antitrust jurisprudence," is *Eastman Kodak Company v. Image Technical Services*. In *Kodak*, the Supreme Court expanded the notion of market power, an element critical to most antitrust violations, to include information. Since agricultural markets are defined by stark information disparities, they present an appropriate context for courts to consider problems such as "searching" for the best price for a product, an information problem identified by George Stigler in the early stages of information economics. One study of Iowa hog farmers, for example, indicates that price searching is very limited and that 85 percent of a farmer's hogs are sold to the same packer, indicating little shopping around. Commentators have noted how "firms can exploit in numerous ways the bargaining power that the lack of comparison shoppers confers on them."[6]

The power-buyer defense to mergers and the recognition of information gaps in certain markets are part of a larger rethinking of antitrust analysis. The greater consideration of complexities in antitrust cases has become known as post-Chicago analysis. Perhaps the most important aspect of post-Chicago analysis is what one commentator describes as the "emergence of sophistication doctrine." Instead of assuming economic rationality among all firms in a market, some recent antitrust cases consider the presence of sophisticated firms that possess "tactical expertise, knowledgeability, or intelligence." The consideration of a firm's sophistication involves an "empirical, improvisational approach to corporate behavior" that allows courts to consider the relative bargaining power between large food-processing firms and small, disorganized farmers. In the case of agribusiness mergers, which often involve large, powerful firms, the sophistication consideration could substantially alter the outcome of antitrust decisions, especially when combined with monopsony considerations.[7]

Courts can also begin to recognize the existence of agricultural statutes in pari materia, which "relate to the same thing" as the antitrust statutes, considering both as "one law" in judicial decision making. Failing to consider agricultural statutes such as the Capper-Volstead Act and the Agricultural Fair Practices Act eliminates critical factors to be considered in antitrust decisions and undermines the designs of legislators. As a broad principle, weighing an array of factors, including closely related statutes, is recognized as an important component of balanced legislative interpretation. If courts consider the wider statutory regime and the problem of farmer disorganization it addressed, judicial decisions can more properly reflect congressional concern about economic concentration and its negative impact on the bargaining power of farmers.[8]

Antitrust law, particularly in recent decades, has failed to consider its agrarian grounding. By incorporating the economic theories of the Chicago school into its analysis, it has failed to take structure as a serious factor in decision making. As a result, noneconomic considerations advanced by Congress such as decentralization have been spurned, contributing to a persistence of economic concentration in certain sectors of the American economy. Consequently, the monopsonistic relationship between some sellers and buyers, a structural consideration of particular importance to farmers, has not been widely recognized by the courts.

In the future, courts should weigh the agrarian origins of the antitrust laws and the importance of structural factors when deciding antitrust cases. In so doing, courts can elaborate on recent developments in antitrust law, mostly outside the agricultural context, which question the usefulness of Chicago analysis. By applying the information analysis of Kodak, courts can take into account the power differential between farmers who lack information about market conditions and large processing firms who have more information than any other entity in the market. The possession of information is also a component of "sophistication" analysis, which does not naively assume an equal footing for market actors, but recognizes that mom and pop often exist within markets alongside a multibillion dollar multinational firm. Such a firm possesses bargaining power over those who sell to it, explaining why some courts allow the merger of large sellers when a "power-buyer" is present in an adjacent market. It also explains why a few courts have considered the existence of monopsony power. The emergence of "post-Chicago" antitrust analysis allows for greater consideration of the particulars in antitrust cases, lending further legitimacy to the analysis of factors such as information availability and the sophistication of firms. Finally, courts can overcome a major oversight in past antitrust cases involving farmers: the failure to consider the range of agricultural statutes and public policies designed to supplement the antitrust laws and bolster the relative bargaining power of the individual farmer, historically disorganized and susceptible to monopsony power.

NOTES

ABBREVIATIONS

DSU	Dakota State University
DWU	Dakota Wesleyan University
GTA	Grain Terminal Association
ISU	Iowa State University
KSHS	Kansas State Historical Society
KSU	Kansas State University
MHS	Minnesota Historical Society
MLPU	Mudd Library, Princeton University
NCFC	National Council Farmer Cooperatives
NFO	National Farmers Organization
NFU	National Farmers Union
NSHS	Nebraska State Historical Society
UO	University of Oklahoma
PU	Purdue University
SDSHS	South Dakota State Historical Society
SDSU	South Dakota State University
UCB	University of Colorado at Boulder
UI	University of Iowa
UND	University of North Dakota
USD	University of South Dakota
UW	University of Washington
WHS	Wisconsin Historical Society

PREFACE

1. J. G. A. Pocock, *Politics, Language, and Time: Essays on Political Thought and History* (New York: Atheneum, 1971), 98, 100.

2. Christopher Lasch, *The True and Only Heaven: Progress and Its Critics* (New York: Norton, 1991), 39.

3. Thomas Bender, "'Venturesome and Cautious': American History in the 1990s," *Journal of American History* 81 (Dec. 1994): 996. See also Jon Lauck, "The History Crisis," *Modern Age* 40, no. 2 (spring 1998): 161–68 and "History without History," *The Social Critic* 3, no. 3 (summer 1998): 14–18.

4. Robert Putnam, "Bowling Alone: America's Declining Social Capital," *Journal of Democracy* 6 (Jan. 1995): 65–78; Michael Sandel, *Democracy's Discontents: America in Search of a Public Philosophy* (Cambridge MA: Harvard University Press, 1996); Christopher Lasch, *The Revolt of the Elites and the Betrayal of Democracy* (New York: Norton, 1995), 19; Arthur Schlesinger Jr., *The Disuniting of America: Reflections on a Multicultural Society* (New York: Norton, 1992).

5. Lincoln quoted in John Patrick Diggins, *The Lost Soul of American Politics: Virtue, Self-Interest, and the Foundations of Liberalism* (New York: Basic Books, 1984), 308–9.

6. Michael Sandel, "America's Search for a New Public Philosophy," *Atlantic Monthly*, Mar. 1996, 69; T. S. Eliot, *Christianity and Culture: The Idea of a Christian Society and Notes Towards the Definition of Culture* (New York: Harcourt, Brace, & World, 1940), 12; Lincoln in Diggins, *Lost Soul of American Politics*, 308–9.

1. THE PROBLEM

1. Galbraith and Bain are quoted in Clair Wilcox, "On the Alleged Ubiquity of Oligopoly," *American Economic Review* 40, no. 2 (May 1950): 67; Mondale is quoted in Steven M. Gillon, *The Democrats' Dilemma: Walter F. Mondale and the Liberal Legacy* (New York: Columbia University Press, 1992), 7; Homer Ayres to Tony Dechant, Mar. 29, 1967, FF 4, DB 53, ser. 3, NFU Papers, UCB.

2. Scott Lash and John Urry, *The End of Organized Capitalism* (Madison: University of Wisconsin Press, 1987), 2; Paul M. Sweezy and Paul A. Baran, *Monopoly Capital* (New York: Monthly Review Press, 1966). The Sweezy quote comes from his article "The Crisis of American Capitalism," *Monthly Review* 22, no. 5 (Oct. 1980): 3; Joseph Schumpeter, *Capitalism, Socialism, and Democracy*, 3d ed. (New York: Harper & Brothers, 1950), 162. For the importance of the idea of corporate power in postwar social science, see, for prominent examples, C. Wright Mills, *The Power Elite* (New York: Oxford University Press, 1957); Gabriel Kolko, *Wealth and Power in America: An Analysis of Social Class and Income Distribution* (New York: Praeger, 1962); G. William Domhoff, *Who Rules America?* (Englewood Cliffs NJ: Prentice-Hall, 1967); John Kenneth Galbraith, *The New Industrial State* (New York: New American Library, 1968); John Bellamy Foster, *The Theory of Monopoly Capitalism: An Elaboration of Marxian Political Economy* (New York:

Monthly Review, 1986); and the debate over corporate control of state decision making in Theda Skocpol, *Bringing the State Back In* (New York: Cambridge University Press, 1985) and G. William Domhoff, *State Autonomy or Class Dominance: Case Studies on Policy Making in America* (New York: Aldine De Gruyter, 1996).

3. Harold Laski quoted in Roger Griffin, ed., *Fascism* (New York: Oxford University Press, 1995), 276; Daniel Guerin, *Fascism and Big Business* (New York: Pathfinder Press, 1973 [1945]), 22; Daniel Bell, "The Power Elite Reconsidered," *American Journal of Sociology* 64 (November 1958), 244 n. 7. The DOJ official is quoted in Morton Mintz and Jerry S. Cohen, *America, Inc.: Who Owns and Operates the United States* (New York: Dial, 1971), 16. Simons quoted in Tony Freyer, *Regulating Big Business: Antitrust in Great Britain and America, 1880–1990* (New York: Cambridge University Press, 1992), 277–78. FDR said in 1938 that "The liberty of a democracy is not safe if the people tolerate the growth of private power to a point where it becomes stronger than their democratic state itself. That, in its essence, is Fascism." Charles S. Maier, *In Search of Stability: Explorations in Historical Political Economy* (New York: Cambridge University Press, 1987), 130–34; Ellis W. Hawley, *The New Deal and the Problem of Monopoly: A Study in Economic Ambivalence* (Princeton NJ: Princeton University Press, 1966), 393, 425; Robert Pitofsky, "The Political Content of Antitrust," *University of Pennsylvania Law Review* 127 (April 1979), 1062–64

4. Marty Strange, "Transforming the Rot Belt," *Des Moines Register*, Feb. 25, 1996; Osha Gray Davidson, *Broken Heartland: The Rise of America's Rural Ghetto* (New York: Free Press, 1990); Mr. Ball to Senator Hubert Humphrey, Senatorial Files, 1971–78, 150.J.2.2(F), DB 1, Hubert Humphrey Papers, MHS; Robert West Howard, *The Vanishing Land* (New York: Villard, 1985), 202; Center for Rural Affairs Annual Report, 1986–87, DB T-179, FF 1/3, Center for Rural Affairs Papers, ISU; James Abourezk, *Congressional Record*, 93d Cong., 1st sess. (Feb. 21, 1973), 119, pt. 4: 4819. These stories are similar to those told about the "loss," "failure," and "collapse" of the U.S. economy in the 1970s. D. N. McCloskey, "1066 and a Wave of Gadgets: The Achievements of British Growth," in Penelope Gouk, ed., *Wellsprings of Achievement: Cultural and Economic Dynamics in Early Modern England and Japan* (Aldershot: Variorum, 1995), 129, and *If You're So Smart: The Narrative of Economic Expertise* (Chicago: University of Chicago Press, 1990), 150–62. For the specific concerns of farmers about the "power of monopolies" see Catherine McNicol Stock, *Rural Radicals: Righteous Rage in the American Grain* (Ithaca NY: Cornell University Press, 1996), 5.

5. Abraham Lincoln, "The Egypt of the West," and Frederick Jackson Turner, "The Middle West," reprinted in *Midwesterner* 2, no. 1 (Feb. 1997), 48, 50; Gilbert

Fite, *American Farmers: The New Minority* (Bloomington: Indiana University Press, 1981), 118; Department of Agriculture Proposal on U.S. Participation in the International Wheat Agreement, DB 5, U.S. Council on Foreign Economic Policy, Policy Paper Series, Eisenhower Library; William C. Pratt, "Using History to Make History? Progressive Farm Organizing during the Farm Revolt of the 1980s," *Annals of Iowa* 55 (winter 1996): 25.

6. Mary Summers, "Putting Populism Back In: Rethinking Agricultural Politics and Policy," *Agricultural History* 70 (spring 1996): 402; Theodore Saloutos and John D. Hicks, *Agricultural Discontent in the Middle-West, 1900–1939* (Madison: University of Wisconsin Press, 1951), 31; Gilbert C. Fite, *George N. Peek and the Fight for Farm Parity* (Norman: University of Oklahoma Press, 1954), 122.

7. On the prevalence of market instability and efforts to reduce farmer uncertainty through such measures as cooperatives see James H. Stock, "Real Estate Mortgages, Foreclosures, and Midwestern Agrarian Unrest, 1865–1920," *Journal of Economic History* 44 (Mar. 1984): 91, and Robert A. McGuire, "Economic Causes of Late-Nineteenth Century Agrarian Unrest: New Evidence," *Journal of Economic History* 41 (Dec. 1981): 837–38.

8. Statement of the National Council of Farmer Cooperatives submitted to the Committee on the Judiciary's Subcommittee on Monopolies and Commercial Law, July 27, 1973, FF 13, DB 17, ser. 3, NFU Papers, UCB; Angus McDonald, Legislative Analysis Memorandum no. 7-67, "A Summary and a More Detailed of the Background and Evolution of S. 109," Oct. 20, 1967, FF 5, DB 1, NFU–NCFC Papers, ISU; James Patton and Angus McDonald, "The Monopoly Squeeze," *Farm Policy Forum* 4 (Jan. 1951): 30.

9. Jeremy Atack and Fred Bateman, "Self-Sufficiency and the Marketable Surplus in the Rural North, 1860," *Agricultural History* 58 (July 1984): 296. I am not ignoring the work of scholars like Winifred Rothenberg and Joyce Appleby who talk about farmers' earlier market participation. Market participation accelerated in the late nineteenth century and became more closely linked to the fortunes of the industrial economy.

10. Richard T. Farrel, "Advice to Farmers: The Content of Agricultural Newspapers, 1860–1910," in Thomas R. Wessel, ed., *Agriculture in the Great Plains, 1876–1936* (Washington DC: Agricultural History Society, 1977), 215; Adam Ward Rome, "American Farmers as Entrepreneurs, 1870–1900," *Agricultural History* 56 (Jan. 1982): 37–49; John T. Schlebecker, *Whereby We Thrive: A History of American Farming, 1607–1972* (Ames: Iowa State University Press, 1975), 160–61.

11. Page Smith, *The Rise of Industrial America: A People's History of the Post-Reconstruction Era*, vol. 6 (New York: McGraw-Hill, 1984), 128; Ellis W. Hawley,

The Great War and the Search for a Modern Order: A History of the American People and Their Institutions, 1917–1933 (New York: St. Martin's, 1992), 5.

12. Alfred D. Chandler Jr. and Richard S. Tedlow, *The Coming of Managerial Capitalism: A Casebook on the History of American Economic Institutions* (Homewood IL: Irwin, 1985), 227; Smith, *Rise of Industrial America*, 128; Robert Higgs, *Crisis and Leviathan: Critical Episodes in the Growth of American Government* (New York: Oxford University Press, 1987), 77–79; Robert F. Lanzillotti, "The Superior Market Power of Food Processing and Agricultural Supply Firms: Its Relation to the Farm Problem," *Journal of Farm Economics* 42 (Dec. 1960): 1228–47.

13. Joyce Appleby, *Capitalism and a New Social Order: The Republican Vision of the 1790s* (New York: New York University Press, 1984), 43; Gary Gerstel, "The Protean Character of American Liberalism," *American Historical Review* 99 (Oct. 1994): 1046; Joyce Appleby, Lynn Hunt, and Margaret Jacob, *Telling the Truth about History* (New York: Norton, 1994), 130–31; Fite, *American Farmers*, 48; Hawley, *New Deal and the Problem of Monopoly*, 293.

14. Joint statement by Wayne Morse and Gaylord Nelson, July 1, 1968, referencing John Kenneth Galbraith's *New Industrial State*, M74-549, DB 69, FF Small Business Committee, Gaylord Nelson Papers, WHS; Statement of Louis B. Schwartz, Senate Small Business Committee, Apr. 27, 1955, Senatorial Files, 1949–64, 150.D.13.1(B), Humphrey Papers, MHS.

15. Fite, *American Farmers*, 128. The term was coined in 1954 by John H. Davis, assistant secretary of agriculture. Most historians agree with Davis. John Shover, *First Majority, Last Minority: The Transforming of Rural Life* (DeKalb: Northern Illinois University Press, 1976), 166; James T. Bonnen, "Observations on the Changing Nature of National Agricultural Policy Decision Processes, 1946–76," in Trudy H. Peterson, ed., *Farmers, Bureaucrats, and Middlemen: Historical Perspectives on American Agriculture*, (Washington DC: Howard University Press, 1980), 312; Trudy Peterson, *Agricultural Exports, Farm Income, and the Eisenhower Administration* (Lincoln: University of Nebraska Press, 1979), 15. Joel Solkoff, in a somewhat different interpretation, says the Earl Butz confirmation hearings in the early 1970s "introduced the public to the little known word 'agribusiness'" (*The Politics of Food* [San Francisco: Sierra Club Books, 1985], 8).

16. Temporary National Economic Committee, *Agriculture and the National Economy*, monograph no. 23 (Washington DC: GPO, 1940), 22; Statement of Russel C. Parker, Senate Small Business Committee, Dec. 10, 1973, DB SenA0056A, James Abourezk Papers, USD; Federal Trade Commission, *The Structure of Food Manufacturing*, technical study no. 8, National Commission on Food Marketing, June 1966, 59; Paul D. Scanlon, "FTC and Phase II: The McGovern Papers," *Anti-*

trust Law and Economics Review 5 (spring 1972): 19–36; Mo Udall asserted that if "food oligopolies were broken up, prices would drop 25%" in a speech to the American Bar Association Conference of Bar Presidents, entered into *Congressional Record* by Rep. Robert Kantenmeir (D, WI), 94th Cong., 1st sess. (Feb. 26, 1975), 121, pt. 4: 4511.

17. A. V. Krebs, *The Corporate Reapers: The Book of Agribusiness* (Washington DC: Essential Books, 1992), 303; Secretary of Agriculture Orville Freeman's quote appeared in the *National Observer*, July 29, 1968; Senator James Abourezk, "Agriculture, Antitrust, and Agribusiness: A Proposal for Federal Action," *South Dakota Law Review* 20 (summer 1975): 499.

18. Hawley, *New Deal and the Problem of Monopoly*, 4; Jim Hightower, Coordinator, Food Action Campaign, to Lewis A. Engman, Chairman, Federal Trade Commission, Nov. 19, 1973, DB SenA0056A, Abourezk Papers, USD; Reverend Eugene L. Boutilier, Director, National Campaign for Agricultural Democracy, to Senator Gaylord Nelson, May 20, 1968, M77-549, DB 255, FF Agriculture-Corporate Farming, Nelson Papers, WHS.

19. Press release, excerpts from remarks of Senator Robert F. Kennedy at Otoe County (NE) Courthouse, May 10, 1968, Kennedy Library; Richard Hofstadter, *The Age of Reform: From Bryan to FDR* (New York: Vintage, 1955): 7; *New York Times* editorial, Dec. 28, 1971; David Lynch, *The Concentration of Economic Power* (New York: Johnson, 1946), 292.

20. The truck driver who gave Tom Joad a lift called it being "tractored out" (John Steinbeck, *The Grapes of Wrath* [New York: Penguin Books, 1976 (1939)], 12); review of *Biting the Dust: The Wild Ride and Dark Romance of the Rodeo Cowboy and the American West*, *New York Times Book Review*, Dec. 25, 1994; Davidson, *Broken Heartland*; Abourezk, *Congressional Record*, 93d Cong., 1st sess., 119, pt. 4: 4819; Robert West Howard, *The Vanishing Land* (New York: Villard, 1985), 202.

21. CRA Annual Report 1982, Center for Rural Affairs Papers, DB T-179, FF 1/3, ISU; Leo Marx, *Machine in the Garden: Technology and the Pastoral Ideal in America* (New York: Oxford University Press, 1964), 6; Wendell Berry, *Home Economics* (New York: North Point, 1987), 174; David B. Danbom, "Romantic Agrarianism in Twentieth-Century America," *Agricultural History* 65 (fall 1991): 9–11.

22. Merril Gilfillan, *Magpie Rising: Sketches from the Great Plains* (New York: Vintage, 1988), 17.

23. Clara B. Riveland, "An Analysis of the National Farmers Organization's Attempts to Reduce Rhetorical Distance" (Ph.D. diss., University of Minnesota, 1974), 44–45.

24. Mrs. Marlowe Carlson to Karl Rolvaag, Sept. 21, 1964, FF Favorable NFO,

110.F.17.14(F), Karl Rolvaag Files, MHS; Matt Shirn to Milton Young, Feb. 22, 1968, and Mrs. Art Hougeberg to Milton Young, Feb. 19, 1968, 20-257-13, Milton Young Papers, UND.

25. Charles A. Stoerzinger, *Current Economic Progress Report for the Upper Midwest, 1964* (Minneapolis: Upper Midwest Research and Development Council, Oct. 1965), 120; Hubert Humphrey, "A Plan for Breathing New Life into Rural America," *Los Angeles Times*, Sept. 26, 1971; Abourezk, *Congressional Record*, 93d Cong., 1st sess., 119, pt. 4: 4819.

26. Bureau of the Census, *Census of Manufacturers, 1947*, vol. 3, *Statistics by States* (Washington DC: GPO, 1950), 567; Bureau of the Census, *Census of Manufacturers, 1992*, Geographic Area Series (Oct. 1995), SD-5; Peter K. Eisinger, *The Rise of the Entrepreneurial State: State and Local Economic Development Policy in the United States* (Madison: University of Wisconsin Press, 1988); Dennis O. Grady, "Governors and Economic Development Policy: The Perception of Their Role and the Reality of Their Influence," *Policy Studies Journal* 17, no. 4 (summer 1989): 879–82; Jim Chen, foreword, "Filburn's Forgotten Footnote: Of Farm Team Federalism and Its Fate," in "Symposium: The Law and Economics of Federalism," *Minnesota Law Review* 82, no. 2 (Dec. 1997): 258–59.

27. William C. Pratt, "Change, Continuity, and Context in Nebraska History, 1940–1960," *Nebraska History* 77, no. 1 (spring 1996): 50; Mark Friedberger, "The Transformation of the Rural Midwest, 1945–1985," *Old Northwest* 16, no. 1 (spring 1992): 23; Eldon Roberts to Truman David Wood, Feb. 14, 1961, in Truman David Wood, "The NFO in Transition" (Ph.D. diss., University of Iowa, 1961), 181; Randy D. Parvin and Steven G. Koven, "Limits on Economic Development Policy: State-Supported Gambling in Iowa," *Policy Studies Review* 14, nos. 3–4 (autumn–winter 1995–96): 431–33; Donald D. Stull, "Rural Industrialization: The Example of Garden City, Kansas," *Kansas Business Review* 14 (1991): 1, and "Cattle Cost Money: Beefpacking's Consequences for Workers and Communities," *High Plains Anthropologist* 14 (1994): 64; Jon Lauck, "The Political Economy of South Dakota Farming, 1945–1995," *Papers of the Twenty-Eighth Annual Dakota History Conference* (The Center for Western Studies and the South Dakota Humanities Council, 1996); Harold Breimyer, *Policies, Attitudes, and Outlook for Economic Development in South Dakota: Highlights of the 10th Agribusiness Day, April 4, 1972*, economic pamphlet 140, Economic Development, Agricultural Extension Station, SDSU, 8. See also Michelle Hoyman, *Power Steering: Global Automakers and the Transformation of Rural Communities* (Lawrence: University Press of Kansas, 1997), and my review in *Annals of Iowa* 57, no. 2 (spring 1998): 189–90.

28. Colin Gordon, *New Deals: Business, Labor, and Politics in America, 1920–1935* (New York: Cambridge University Press, 1994), 1–2, 31, 33, 139.

29. Jeremy Atack and Peter Passell, *A New Economic View of American History: From Colonial Times to 1940*, 2d ed. (New York: Norton, 1994), 462–64, 487–88; Michael J. McGarry and Andrew Schmitz, *The World Grain Trade: Grain Marketing, Institutions, and Policies* (London: Pinter, 1992), 488; Willard Cochrane and Mary Ryan, *American Farm Policy, 1948–1973* (Minneapolis: University of Minnesota Press, 1976), 10–12; Fite, *American Farmers*, 177.

30. For examples of the conventional wisdom about oligopoly pricing in the food processing sector see Lanzillotti, "Superior Market Power," 1232, 1240, and 1244; statement of Russel C. Parker, Select Committee on Small Business Hearings, Monopoly Subcommittee, Dec. 10, 1973, 3, DB SenA0056A, Abourezk Papers, USD; and Shover, *First Majority, Last Minority*, 166–67, 188; Robert Bork, *The Antitrust Paradox: A Policy at War with Itself* (New York: Basic Books, 1978), 178–97; Richard Posner, "Oligopoly and the Antitrust Laws: A Suggested Approach," *Stanford Law Review* 21 (June 1969): 1566–67. For more background on the debate and its implications for antitrust policy see George A. Hay, "Oligopoly, Shared Monopoly, and Antitrust Law," *Cornell Law Review* 67 (Mar. 1982): 439–81.

31. D. N. McCloskey, "The Economics of Choice: Neoclassical Supply and Demand," in *Economics and the Historian* (Berkeley and Los Angeles: University of California Press, 1996), 133, "The Arrogance of Economic Theorists," *Swiss Review of World Affairs* 41 (Oct. 1991): 12, "Does the Past Have Useful Economics?" *Journal of Economic Literature* 14 (June 1976): 434, and *The Vices of Economists, The Virtues of the Bourgeoisie* (Amsterdam: Amsterdam University Press, 1996), 123–24.

32. Alan Brinkley, "Writing the History of Contemporary America: Dilemmas and Challenges," *Daedalus* 113 (summer 1984): 139; Allan Megill and D. N. McCloskey, "The Rhetoric of History," in John Nelson, Allan Megill, and D. N. McCloskey, eds., *The Rhetoric of the Human Sciences: Language and Argument in Scholarship and Public Affairs* (Madison: University of Wisconsin Press, 1987), 233.

33. Frederick Scherer, "The Posnerian Harvest: Separating Wheat from Chaff," review of Richard Posner's *Antitrust Law: An Economic Perspective* in *Yale Law Journal* 86 (Apr. 1977): 995. Scherer agrees with Posner that "price competition in concentrated industries may be a good deal more vigorous than one might anticipate on the basis of naive oligopoly theories."

34. Willard F. Mueller, "Empirical Measurement in Market Structure Research," *Journal of Farm Economics* 43 (Dec. 1961): 1372.

35. Robert L. Clodius and Willard F. Mueller, "Market Structure as an Orientation for Research in Agricultural Economics," *Journal of Farm Economics* 43 (Aug. 1961): 517 n. 3.

36. Lanzillotti, "Superior Market Power," 1229.

37. Statement of Russel C. Parker. Clodius and Mueller noted that "all aspects of market structure analysis rest on the basic assumption that market structure determines, in large part, the competitive conduct of firms in a market which in turn generates certain forms of industrial performance" ("Market Structure as an Orientation," 529). Donald J. Dewey doubts the validity of the theories stemming from the "simpler intellectual world of Frank Fetter, Justice Brandeis, William Ripley, and Henry Simons" in his essay "The New Learning: One Man's View" in Harvey J. Goldschmid, H. Michael Mann, and J. Fred Watson, eds., *Industrial Concentration: The New Learning* (Boston: Little, Brown, 1974), 8. The book is considered the "opening shot" of the new debate over the causes and consequences of concentration. The book ignores the question of the relative organizational power of the farm sector.

38. Report of the White House Task Force on Antitrust Policy, *Antitrust Law and Economics Review* 2 (winter 1968–69): 12.

39. Bork, *Antitrust Paradox*, 217. From 1950 to 1977 the Department of Justice brought 135 cases against the food manufacturing sector and the Federal Trade Commission lodged 256 complaints; Bruce W. Marion, "Government Regulation of Competition in the Food Industry," *American Journal of Agricultural Economics* 61 (Feb. 1979): 179–80. In the 1960s the Warren Court used market share percentages to adjudicate antitrust cases: "We think that a merger which produces a firm controlling an undue percentage share of the relevant market, and results in a significant increase in the concentration of firms in that market, is so inherently likely to lessen competition substantially that it must be enjoined in the absence of evidence clearly showing that the merger is not likely to have such anti-competitive effects" (*United States v Philadelphia National Bank et al.*, 374 U.S. 321, 363, quoted in Stephen J. Heimstra, "Concentration and Competition in the Food Industries," *Journal of Farm Economics* 48 [Aug. 1966]: 138). Doubts about the Bain school of thought don't mean that concentration isn't linked to market power and high prices. Sometimes it is. The point is that the connection is not obvious. For scattered studies indicating that the older theories still hold water see Bruce W. Marion, "Interrelationships of Market Structure, Competitive Behavior, and Market/Firm Performance: The State of Knowledge and Some Research Opportunities," *Agribusiness* 2 (1986): 449–50. In the end, "very little systematic evidence has been gathered on the history of collusive agreements and, while casual observation abounds (antitrust practitioners are familiar with some examples of collusion that seem to work smoothly, others that appear to break down), this kind of information, because of its unfortunate tendency to support virtually any conclusion, is of rather limited use to the policymaker" (Peter Asch, "Collusive Oligop-

oly: An Antitrust Quandary," *Antitrust Law and Economics Review* 2 [spring 1969]: 65). For the historical ambiguities in industrial organization, see Herb Hovenkamp, *Enterprise and American Law, 1836–1937* (Cambridge MA: Harvard University Press, 1991), 296–307.

40. P. David Qualls, Assistant Director for Industry Analysis, memo to the FTC, July 22, 1977, FF 18, DB 22, Hart Papers, UCB.

41. D. N. McCloskey, *The Applied Theory of Price* (New York: Macmillan, 1982), 416, and "Does the Past Have Useful Economics?" 443; R. H. Coase, *The Firm, the Market, and the Law* (Chicago: University of Chicago Press, 1988), 61; F. M. Scherer, *Industrial Market Structure and Economic Performance* (Chicago: Rand McNally, 1970), 151.

42. Scherer, *Industrial Market Structure*, 155.

43. John Bellamy Foster and Henryk Szlajfer, introduction, *The Faltering Economy: The Problem of Accumulation under Monopoly Capitalism* (New York, Monthly Review, 1984), 7; Richard Rorty, "The End of Leninism and History as Comic Frame," in Arthur M. Melzer, Jerry Weinberger, and M. Richard Zinman, eds., *History and the Idea of Progress* (Ithaca NY: Cornell University Press, 1995), 211.

44. Scherer, *Industrial Market Structure and Economic Performance* 169–70, 173, 182. Leonard W. Weiss reviewed forty-six concentration/profit studies and found that the "bulk of these studies yielded significant positive relationships between concentration and profits or price-cost margins"; "The Structure-Conduct-Performance Paradigm and Antitrust," *Cornell Law Review* 127 (1979): 1106. William E. Kovacic, "The Detection and Punishment of Tacit Collusion," *Loyola Consumer Law Reporter* 9 (1997): 154; Angela Wissman, "ADM Execs Nailed on Price-Fixing, May Do Time: Government Gets Watershed Convictions, but Company Still Dominates Lysine Market," *Merrill's Illinois Legal Times*, October 1998, 1.

45. Derek C. Bok, "Section 7 of the Clayton Act and the Merging of Law and Economics," *Harvard Law Review* 74(1960): 349.

46. *Eastman Kodak Co. v. Image Technical Services*, 504 U.S. 451, 479 (1992).

47. Robert Pitofsky, "The Political Content of Antitrust," *University of Pennsylvania Law Review* 127 (1979): 1065.

48. Frederick M. Rowe, "The Decline of Antitrust and the Delusions of Models: The Faustian Pact of Law and Economics," *Georgetown Law Journal* 72 (1982): 1513, 1520, 1522, 1569.

49. William O. Douglas, *The Court Years, 1935–1975: The Autobiography of William O. Douglas* (New York: Random House, 1980), 162.

50. George J. Stigler, "A Theory of Oligopoly," *Journal of Political Economy* 72 (Feb. 1964): 44–61.

51. G. E. Brandow, "Market Power and Its Sources in the Food Industry," *American Journal of Farm Economics* 51 (Feb. 1969): 2, 8. John Connor, Dale Heien, Jean Kinsey, and Robert Wills, "Economic Forces Shaping the Food-Processing Industry," *American Journal of Agricultural Economics* 67 (Dec. 1985): 1138–39, focuses only on demand, costs, and structure. In a footnote to the section on structure the authors say that they have focused only on selling strategies and have omitted "procurement practices in input markets." This reflects the lack of importance, or maybe the improbability, assigned to the organization of the input sector, particularly since the biggest expense of food processors are raw farm products that lack substitutes.

52. Hawley, *New Deal and the Problem of Monopoly*, 187–97; John Kenneth Galbraith, *American Capitalism: The Concept of Countervailing Power* (Boston: Houghton Mifflin, 1952), 110–11. Walter Adams voiced several doubts about Galbraith's idea in a 1953 article that is reprinted in James W. Brock and Kenneth Elzinga, eds., *Antitrust, the Market, and the State: The Contributions of Walter Adams* (Armonk NY: Sharpe, 1991), 238–54.

53. Willard Mueller to John Chernauskas, Sept. 6, 1979, FF 28, DB 22, ser. 3, NFU Papers, UCB; John M. Connor, Richard T. Rogers, Bruce Marion, and Willard Mueller, *The Food Manufacturing Industries: Structure, Strategies, Performance, and Policies* (Lexington MA: Heath, 1985), 102; John R. Moore, "Bargaining Power Potential in Agriculture," *American Journal of Agricultural Economics* 50 (Nov. 1968): 1051–53.

54. John H. Davis and Kenneth Hinshaw, *Farmer in a Business Suit* (New York: Simon & Schuster, 1957); Bob Bergland to the National Commission for the Review of the Antitrust Laws and Procedures, July 27, 1978, FF 18, DB 9, ser. 2, NFU Papers, UCB; Gordon Wood, "The Enemy Is Us: Democratic Capitalism in the Early Republic," *Journal of the Early Republic* 16 (summer 1996): 297; Don Paarlberg, "The Land Grant Colleges and the Structure Issue," *American Journal of Agricultural Economics* 63 (Feb. 1981): 129.

55. William Jennings Bryan, "Cross of Gold," in Glenn R. Capp, *Famous Speeches in American History* (Indianapolis: Bobbs-Merrill, 1963), 124.

56. "Proposal for Parallel Action Clearing House," DB 3, FF 29, ser. 3, NFU Papers, UCB; *Business Week*, Dec. 10, 1955, 113, 115; William P. Browne, *Cultivating Congress: Constituents, Issues, and Interests in Agricultural Policymaking* (Lawrence: University Press of Kansas, 1995), 25

57. E. Dale Odom, "Associated Milk Producers, Incorporated: Testing the Limits of Capper-Volstead," *Agricultural History* 59 (Jan. 1985): 40–55; *Wall Street Journal*, Oct. 15, 1973; press release, National Council of Farmer Cooperatives, Feb. 12, 1976, DB 47, FF Judiciary: Food Industry, John Culver Papers, UI.

58. Jackson quoted in Alan Brinkley, *The End of Reform: New Deal Liberalism in Recession and War* (New York: Vintage, 1996 [1995]), 60. Hand quoted in Brock and Elzinga, *Antitrust, the Market, and the State*, 149.

59. Stephen J. Spingarn to Charles S. Murphy, Sept. 26, 1951, Harry S. Truman Official File, Truman Library. Nixon quoted in Neil Fligstein, *The Transformation of Corporate Control* (Cambridge MA: Harvard University Press, 1990), 209.

60. Angus McDonald to Tony Dechant, Dec. 9, 1968, FF 21, DB 5, ser. 3, NFU Papers, UCB.

61. Ben Radcliffe, "Effects of Corporate Farming on Small Business," hearings before the Subcommittee on Monopoly of the Senate Small Business Committee, May 20, 1968, 21; A. Whitney Griswold, *Farming and Democracy* (New York: Harcourt, Brace, 1948), 30.

62. Donald Dewey, *The Antitrust Experiment in America* (New York: Columbia University Press, 1990), 11, 21, 40; Herbert Hovenkamp, "Antitrust Policy after Chicago," *Michigan Law Review* 84 (Nov. 1985): 244–49; Jane B. Baron and Jeffrey L. Dunhoff, "Against Market Rationality: Moral Critiques of Economic Analysis in Legal Theory," *Cardozo Law Review* 17 (Jan. 1996): 432; E. Thomas Sullivan, "The Economic Jurisprudence of the Burger Court's Antitrust Policy: The First Thirteen Years," *Notre Dame Law Review* 58 (Oct. 1982): 57; Michael O. Wise, "Antitrust's Newest 'New Learning' Returns the Law to Its Roots: Chaos and Adaptation as New Metaphors for Competition Policy," *Antitrust Bulletin* 40 (winter 1995): 715–16.

63. Deirdre McCloskey, "Economic Tourism," *Eastern Economic Journal* 22 (summer 1996): 367.

64. Irwin M. Stelzer, "A Conservative Case for Regulation," *Public Interest* no. 128 (summer 1997): 86–87; Jon Lauck, "Reviving Republicanism," *The Social Critic* 3, no. 1 (winter 1998): 8; Federalist 10, *The Federalist Papers* (New York: New American Library, 1961), 77.

65. Hayden White, *Tropics of Discourse: Essays in Cultural Criticism* (Baltimore: Johns Hopkins University Press, 1978), 46.

66. Hofstadter quoted in Christopher Lasch, "Consensus: An Academic Question," *Journal of American History* 76 (Sept. 1989): 458.

67. William E. Leuchtenburg, "The Pertinence of Political History: Reflections on the Significance of the State in America," *Journal of American History* 73 (Dec. 1986): 585; Robert M. Collins, "The Economic Crisis of 1968 and the Waning of the 'American Century,'" *American Historical Review* 101 (Apr. 1996): 398. Eugene Genovese, Elizabeth Fox-Genovese, and their critics are quoted in Peter Novick, *The Noble Dream: The "Objectivity Question" and the American Historical Profession* (New York: Cambridge University Press, 1988), 443; Call for Abstracts, H-Net Humanities Graduate Discussion List, Mar. 17, 1997.

68. Jean Bethke Elshtain, *Democracy on Trial* (New York: Basic Books, 1995), 20–21; Todd Gitlin, *The Twilight of Common Dreams: Why America Is Wracked by Culture Wars* (New York: Metropolitan Books, 1995), 3; Lasch, *Revolt of the Elites*, 3; Michael Sandel, *Democracy's Discontent: Americans in Search of a Public Philosophy* (Cambridge MA: Harvard University Press, 1996). Beveridge quoted in Leuchtenberg, "The Pertinence of Political History," 590.

2. THE CORPORATE FARMING DEBATE

Dan Turner was governor in the 1930s but continued his farm activism into the 1950s.

1. Douglas Unger, *Leaving the Land* (Lincoln: University of Nebraska Press, 1984), 73–74, 153–54, 176–79.

2. Hiram M. Drache, "Midwest Agriculture: Changing with Technology," *Agricultural History* 50 (Jan. 1976): 291; Philip M. Raup, "Corporate Farming in Agriculture," *Journal of Economic History* 33 (Mar. 1973): 274, 279; *New Land Review* 1 (fall 1974): DB T-179, FF 1/7, Agribusiness Accountability Project Papers, ISU.

3. Text of Philip M. Raup, "Needed Research into the Effects of Large Scale Farm and Business Firms on Rural America," Mar. 1, 1972, DB SA0056C, Abourezk Papers, USD; M. W. Thatcher to Farmers Union rally, Pipestone MN, Apr. 1953, FF Speeches by Thatcher, 1953–1954, DB 2, 149.E.10.7(B), M. W. Thatcher Papers, MHS; Angus McDonald, "Vertical Integration Trends: Particularly in Regard to the National Tea Company," Supplement 1, Legislative Analysis Memorandum 58-8, Dec. 11, 1963, NFU–NCFC Papers, ISU.

4. Neil F. Harl papers "Influencing the Structure of Agriculture," FF Farm Structure, DB Gilbert Agricultural Files 213 and "Corporate S AND 1977 Farm Legislation," FF Corporate Farming, DB Gilbert Agricultural Files 200, Culver Papers, UI.

5. *Omaha World-Herald*, June 18, 1996; *National Journal*, Mar. 12, June 18, 1983; *Dakota Farmer*, June 1994; *Time*, Mar. 18, 1996; *Des Moines Register*, May 5, 1996.

6. Fite, *American Farmers*, 127; Sydney Gross (President of the Iowa Farmers Union) and Ben Radcliffe (President of the South Dakota Farmers Union), "Effects of Corporation Farming on Small Business," Hearings before the Subcommittee on Monopoly of the Senate Small Business Committee, May 20, 1968 (Washington DC: GPO, 1968), 16, 18.

7. Oren Lee Stately, copy of statement to Senate Small Business Committee, Dec. 1, 1971, DB A0056B food hearings, Abourezk Papers, USD; *NFO Reporter*, Nov. 1971.

8. *Washington Post*, Oct. 5, 1971; Morton Rothstein, "The Big Farm: Abundance and Scale in American Agriculture," *Agricultural History* 49 (Oct. 1975): 586; Breimyer, *Policies, Attitudes, and Outlook*, 4.

9. Agribusiness Accountability Project, "Business as Usual: Corporate Influence in Food Policy," Dec. 1973, Agribusiness Accountability Project Papers, ISU; Gaylord Nelson to Gene Hausner, Jan. 12, 1968, FF Agriculture Price Bargaining, Nelson Papers, WHS; Food Action Campaign, "Corporate Concentration in the Food Economy: What Can Be Done?" Abourezk Papers, USD; Gaylord Nelson to Mr. and Mrs. Rudy Kugel, Nov. 21, 1968, FF Agriculture: Corporate Farming, Nelson Papers, WHS.

10. *Washington Post*, Oct. 4, 1971; Wood, "The Enemy Is Us," 305; James Rhodes and Leonard R. Kyle, "A Corporate Agriculture," no. 3 of *Who Will Control U.S. Agriculture?* (North Central Public Policy Education Committee, Extension Service, USDA, and Farm Foundation, n.d.); *Des Moines Register* clipping, c. Feb. 1974.

11. David Obey, *Congressional Record*, 93d Cong., 2d sess. (Feb. 19, 1974), 120, pt. 3: 3512; Nellis quoted in "The Farmers—Again," *Progressive* 43 (Apr. 1979). For the comparison between the food and oil industry see Jim Hightower, "Eat It! Corporate Giantism and Food Profits," *Win* (Jan. 31, 1974), 4; Abourezk press release, Dec. 10, 1973, DB SenA0056A, Abourezk Papers, USD; John Hart, "How Agribusiness Is Destroying Agriculture," *Christianity and Crisis* (Apr. 15, 1985), 130.

12. Patrick J. Akard, "Corporate Mobilization and Political Power: The Transformation of U.S. Economic Policy in the 1970s," *American Sociological Review* 57 (Oct. 1992): 601–2; *Harvard Business Review* is quoted, but not footnoted, in Alvin Toffler, *The Third Wave* (New York: Bantam, 1980), 233; Senators Gaylord Nelson and Wayne Morse, Senate Small Business Committee press release, June 28, 1968, and National Federation of Independent Business clipping, Mar. 1966, both in M74-549, DB 69, FF Small Business Committee, Nelson Papers, WHS; *Brawley News* (California), Dec. 19, 1972; Mo Udall, "The Future of Antitrust," speech to the American Bar Association, Feb. 21, 1975, reprinted in *Congressional Record*, 94th Cong., 2d sess. (Feb. 26, 1975), 121, pt. 30.

13. Robert A. Pastor, *Congress and the Politics of U.S. Foreign Economic Policy* (Berkeley and Los Angeles: University of California Press, 1990), 133–34; Mira Wilkins, "America and the World Economy," in Robert Bremner, Gary W. Reichard, and Richard J. Hopkins, eds., *American Choices: Social Dilemmas and Public Policy since 1960* (Columbus: Ohio State University Press, 1986), 233–34; Thatcher speech text, FF Biographical Information, DB 1, 149.E.10.6(F), Thatcher Papers, MHS.

14. "Agribusiness Corporations Served by the U.S. Secretary of Agriculture

Nominee, Dr. Earl Butz," Report of the Agribusiness Accountability Project, Nov. 17, 1971, FF 8, DB 1, Agribusiness Accountability Project Papers, ISU; *St. Paul Pioneer Press*, Nov. 19, 1971; NFO *County Progress Reporter*, Dec. 1971; *New York Times*, Dec. 28, 1971; Ralph Nader to Mike Mansfield, Nov. 18, 1971, and Ed Wimmer to Richard Nixon, Nov. 16, 1971, FF 25, DB 11, ser. 3, NFU Papers, UCB; Tony Dechant speech to GTA stockholders meeting, Dec. 1, 1971, FF 24, DB 3, ser. 2, NFU Papers, UCB; FF Confirmation Hearings, Senate Agriculture Committee, 1971, DB G, Earl Butz Papers, PU.

15. Abourezk statement to Senate Subcommittee on Monopoly, Dec. 10, 1973, DB SenA0056A, Abourezk Papers, USD.

16. Abourezk press release, June 24, 1972, DB A0056A, Abourezk Papers, USD; McGovern speech to the "People's Dinner," 62; NFO *Reporter*, Feb. 1976; press release, Dec. 1, 1970, FF 2, DB 1, Agribusiness Accountability Project Papers, ISU; Kennedy speech to Iowa Farmers Union Convention, Sept. 9, 1972, FF 46, DB 3, ser. 2, NFU Papers, UCB; Alice and Sam Beattie to Hubert Humphrey, Nov. 29, 1971, 150.J.2.2(F), DB 1, Humphrey Senatorial Files, 1971–78, Humphrey Papers, MHS; Fred Harris to Mr. and Mrs. Sasser, May 25, 1972, FF 35, DB 259, Fred Harris Papers, Albert Center, UO.

17. Jim Hightower, "A Summary of Hard Tomatoes, Hard Times: The Failure of the Land Grant College Complex," Preliminary Report of the Task Force on the Land Grant College Complex, released by Agribusiness Accountability Project, May 31, 1972, DB A0056, Abourezk Papers, USD; Abourezk press release, Dec. 10, 1973, DB SenA0056A, Abourezk Papers, USD; NFO *County Progress Reporter*, June 1972; William P. Browne, "Challenging Industrialization: The Rekindling of Agrarian Protest in a Modern Agriculture, 1977–1987," *Studies in American Political Development* 7 (spring 1993): 23. For Secretary Bergland's response to such criticism see Bergland, "The Federal Role in Agricultural Research," address to the USDA's Science and Education Administration, Jan. 31, 1980, 151.H.11.6(F), DB 2, Bergland Papers, MHS.

18. Nelson press release, Oct. 19, 1979, FF Agriculture and Family Farms, DB 190, M80-626, Nelson Papers, WHS.

19. Robert Schurman to McGovern, May 20, 1959, McGovern to Robert Schurman, May 23, 1959, press release, "Poultry Hearing Set," Apr. 21, 1959, McGovern to Mr. and Mrs. George H. Heibult, Sept. 24, 1959, all in FF 1959 Re Agriculture: Egg and Poultry, DB 1959, Correspondence Re Legislation: Agriculture Aa-Rz, George McGovern Papers, MLPU.

20. *Washington Post*, June 10, 1960; McGovern speech, Des Moines, Oct. 6, 1972, in McGovern, *American Journey*, 63; copy of "Farming and Rural Life" speech in FF Gen from Platform and DCD Reorganization, DB SenA0056,

Abourezk Papers, USD; McGovern speech, Sioux Falls, Sept. 24, 1972, FF 13, DB 25, ser. 3, NFU Papers, UCB; Brinkley, *End of Reform*, 5, 7, 58–60.

21. Testimony reprinted in the *Congressional Record Appendix*, 86th Cong., 1st sess. (Feb. 18, 1959), 105: A1212; Hart, "How Agribusiness Is Destroying Agriculture," 134; *Center for Rural Affairs Newsletter* 1, no. 5 (Dec. 1973), Center for Rural Affairs Papers, ISU; *Land Reform*, World Bank Paper, Rural Development Series (July 1974), 20.

22. News release, National Catholic Rural Life Conference Executive Committee, Jan. 19, 1967, FF 4, DB 3, ser. 3, NFU Papers, UCB; Bishop Dingman, *Catholic Rural Life* 28, no. 1 (Jan. 1979): 2, 23. See also the article about Bishop Speltz's testimony to Congress in *Catholic Rural Life* 28, no. 4 (Apr. 1979): 15; NFO *County Progress Reporter*, Apr. 1979.

23. Edward W. O'Rourke, "Urges Farmers to Try Collective Bargaining," *Catholic Messenger*, Dec. 29, 1960; James Patton to Edward O'Rourke, Oct. 6, 1964, FF 16, DB 7, ser. 4, NFU Papers, UCB; Monsignor Louis J. Miller, talk delivered over KELO radio, Mar. 1967, text in FF 8, DB 7, NFO Papers, ISU; Willis Rowell, *Mad as Hell: A Behind the Scenes Story of the NFO* (Corning IA: Gauthier, 1984), 18; Denton E. Morrison and Allan D. Steeves, "Deprivation, Discontent, and Social Movement Participation: Evidence on a Contemporary Farmers' Movement, The NFO," *Rural Sociology* 32, no. 4 (Dec. 1967): 422; Walter E. Carlson to Milton Young, Apr. 14, 1967, FF 20-257-13, Young Papers, UND.

24. Michael Perelman and Kevin P. Shea, "The Big Farm," *Environment* 14 (Dec. 1972): 10–14; Senator Nelson and Tony Dechant (President of the National Farmers Union), "Effects of Corporation Farming on Small Business," Hearings before the Subcommittee on Monopoly of the Senate Small Business Committee, May 20, 1968 (Washington DC: GPO, 1968), 11–12, 17; Ron Way to Steve Pavich, Mar. 22, 1972, FF Agriculture, General, DB 346, M80–626, Nelson Papers, WHS.

25. David Danbom, "Romantic Agrarianism in Twentieth-Century America," *Agricultural History* 65 (fall 1991): 10. Butz's review of Berry's book can be found in the actual book in DB F, Butz Papers, PU; *Center for Rural Affairs Annual Report, 1980*, Center for Rural Affairs Papers, ISU; William P. Browne, "Challenging Industrialization: The Rekindling of Agrarian Protest in a Modern Agriculture, 1977–1987," *Studies in American Political Development* 7 (spring 1993): 22.

26. Transcript of the program inserted in *Congressional Record* by Senator Stevenson of Illinois on Sept. 10, 1971, S14081-85; Jerry Berman, Woody Ginsburg, Martha Hamilton, Jim Hightower, Nancy Mills to Reuven Frank, Aug. 18, 1971, FF 7, DB 1, Agribusiness Accountability Project Papers, ISU. In Mar. 1972 the PBS program "The Advocates" also hosted a debate on whether Congress should prohibit corporate farming.

27. *New York Times*, Dec. 5, 1971; *Washington Post*, Oct. 19, 1994; Don Paarlberg, *Corporate Farming and the Family Farm* (Ames: Iowa State University Press, 1970), 115; Neil Harl, "Corporate Farms and 1977 Farm Legislation," 8, FF Corporate Farming, DB Gilbert Agricultural Files 200, Culver Papers, UI.

28. *Washington Post*, Oct. 5, 1971; *Des Moines Register*, Nov. 24, 1971; "Contract Farming and Vertical Integration in Agriculture," Agriculture Information Bulletin no. 198, July 1958, USDA, 5, 17–18; Tony Dechant, "Effects of Corporation Farming on Small Business," Hearings before the Subcommittee on Monopoly of the Senate Small Business Committee, May 20, 1968 (Washington DC. GPO, 1968), 11.

29. Oren Lee Stately, copy of statement to Senate Small Business Committee, Dec. 1, 1971, DB A0056B food hearings, Abourezk Papers, USD; Fite, *American Farmers*, 128; James G. Patton, *The Case for Farmers* (Washington DC: Public Affairs Newsletter, 1959), 42.

30. Jim Hightower, "Corporate Power in Rural America," FF 28, DB 219, Harris papers, Albert Center, UO; *NFO Reporter*, Dec. 1971.

31. *Washington Post*, Oct. 5, 1971; Agribusiness Accountability Project press release, Dec. 1, 1970, FF 2, DB 1, Agribusiness Accountability Project Papers, ISU; Reverend Eugene L. Boutilier (Director of National Campaign for Agricultural Democracy) to Gaylord Nelson, May 20, 1968, FF Agriculture, Corporate Farming, DB M74-549 255, Nelson Papers, WHS.

32. *Center for Rural Affairs Newsletter*, Feb. 1975, Center for Rural Affairs Papers, ISU; "Small Business and the Community: The Effects of the Scale of Farm Operations, Dec. 23, 1946," reprinted in Hearings before the Subcommittee on Monopoly of the Select Committee on Small Business, U.S. Senate, 90th Cong., 2d sess. (Washington DC: GPO, 1968), 3–6; Charles Walters Jr., *Holding Action* (New York: Halcyon House, 1968), xiii. For another example, see Gregory Michaels and Gerald Marousek, "Economic Impact of Farm Size Alternatives on Rural Communities," Bulletin no. 582, May 1978, Agricultural Experiment Station, College of Agriculture, University of Idaho.

33. Russell King, *Land Reform: A World Survey* (Boulder CO: Westview, 1977), 45; *Report of the World Land Reform Conference* (New York: United Nations, 1968), 3; Charles C. Geisler, "A History of Land Reform in the United States," in Charles C. Geisler and Frank J. Popper, *Land Reform, American Style* (Totowa NJ: Rowman & Allanheld, 1984), 24–25; *Feedstuffs*, Mar. 5, 1973; Mark Friedberger, *Shake-Out: Iowa Farm Families in the 1980s* (Lexington: University Press of Kentucky, 1989), 156; Harold F. Breimyer, "Future Organization and Control of U.S. Agricultural Production and Marketing," *Journal of Farm Economics* 47 (Dec. 1964): 939; Harold F. Breimyer, "Farms, Farmers, and Farm Policy in an Industrial

Age," speech in Moorhead MN, Jan. 16, 1962, mailed to author. Breimyer was one of the first economists to tackle the corporate farming issue. Breimyer to author, Sept. 20, 1996; Harold Breimyer, *Over-fulfilled Expectations: A Life and an Era in Rural America* (Ames: Iowa State University Press, 1991), 48.

34. George McGovern, "New Light on the Family Farm," *Congressional Record*, 93d Cong., 1st sess. (May 31, 1973), 119, pt. 82: 17616-25; Fred Harris statement, July 12, 1971, FF 26, DB 256, Harris Collection, Albert Center, UO; Land Conference schedule, MSC 414, FF Land Reform Conference, DB 27, Clark Papers, UI; Tony Dechant to State Farmers Union Presidents, Nov. 1, 1973, FF 46, DB 16, ser. 3, NFU Papers, UCB; Sheldon Greene to Tony Dechant, Dec. 9, 1971, FF 38, DB 14, ser. 3, NFU Papers, UCB; Peter Barnes (National Coalition for Land Reform), "Land Reform in America," in Richard Merrill, ed., *Radical Agriculture* (New York: Harper & Row, 1976), 26-38; "Corporate Invasion in Land Ownership," working paper, Third National Conference on Rural America, Dec. 1977, FF 9, DB 43, ser. 3, NFU Papers, UCB.

35. James Patton, speech to National Democratic Leaders at Des Moines, May 1964, FF Democratic National Convention, DB 67A1881, carton 4, McGovern Papers, MLPU; *NFO Reporter*, Aug. 1975; Glen J. Vollmar, "Factory Farms versus Family Farms," in *Corporation Farming: What Are the Issues?* Department of Agricultural Economics Report no. 53, Proceedings of the North Central Workshop, Apr. 1969, 33. For a three-part series on the need for land reform, see *New Republic* 164, nos. 23, 24, 25 (June 5, 12, 19, 1971).

36. Sydney Gross (President of the Iowa Farmers Union) and McGovern are quoted in the "Effects of Corporation Farming on Small Business," Hearings before the Subcommittee on Monopoly of the Senate Small Business Committee, May 20, 1968 (Washington DC: GPO, 1968), 16, 18.

37. Webster quoted in "Small Business and the Community: The Effects of the Scale of Farm Operations, December 23, 1946," Report of the Special Committee to Study Problems of American Small Business, U.S. Senate.

38. Jefferson quoted in Arthur Schlesinger Jr., *The Age of Jackson* (Boston: Little, Brown, 1946), 8; Victor and Evelyn Matehs to Milton Young, Feb. 24, 1968, FF 20-257-13, Young papers, UND; Tony Dechant speech to Professional Agricultural Workers of Texas Conference, Fort Worth, Aug. 21, 1969, FF 15, DB 3, ser. 2, NFU Papers, UCB; M. W. Thatcher to Farmers Union rally, Pipestone MN.

39. Gaylord Nelson to Carl T. Curtis, Apr. 26, 1968, DB M74-549 255, FF Agriculture, Corporate Farming, Nelson Papers, WHS; Roger Blobaum to Gaylord Nelson, Feb. 8, 1968, FF 47, DB 5, ser. 3, NFU Papers, UCB.

40. *Guardian*, June 1968; *National Farmers Union Washington Newsletter*, May 23, 1968. Gregory Stephens of the Kansas NFO Oral History Project told me

in a telephone conversation on Aug. 20, 1996, that the NFO was one of Kennedy's greatest supporters. Kennedy won eighty of the eighty-six rural counties in Nebraska, for example. *Prairie Farmer*, June 1, 1968; Nelson press release, June 6, 1968, FF Corporate Farming Omaha Witnesses and Eau Claire, DB 12, M74-549, Nelson Papers, WHS.

41. Tony T. Dechant to Gaylord Nelson, Sept. 10, 1968, DB M74-549 255, FF Agriculture, Corporate Farming, Nelson Papers, WHS; Randall E. Torgerson, *Producer Power at the Bargaining Table: A Case Study of the Legislative Life of S. 109* (Columbia: University of Missouri Press, 1970), 232.

42. Manuscript of Victor Ray's "They're Destroying Our Small Towns," May 1, 1974, FF 15, DB 2, ser. 2, NFU Papers, UCB; *Doane's Agricultural Report*, May 1970, 24–25; *Wall Street Journal*, Oct. 15, 1973; *Courier-Journal*, Dec. 14, 1973; David William Seckler, "Why Corporate Farming," 1969, Agricultural Economics Report, College of Agriculture and Home Economics, University of Nebraska–Lincoln; Introduction, "Impact of Corporation Farming on Small Business," Report of the Senate Select Committee on Small Business, 91st Cong., 1st sess., 1968, Report no. 91-628, 1; Tony Dechant, "Effects of Corporation Farming on Small Business," Hearings before the Subcommittee on Monopoly of the Senate Small Business Committee, May 20, 1968 (Washington DC. GPO, 1968), 12; *NFO Reporter*, Aug. 1974, Dec. 1973; statement of James McHale to Subcommittee on Monopoly, Dec. 12, 1973, DB A0056A, Abourezk Papers, USD; Charles Walters, *Angry Testament* (Kansas City: Halcyon House, 1969), 357–61; *Washington Post*, Oct. 3, 1971.

43. *Washington Post*, Oct. 4, 1971; NFO news release, Oct. 1, 1971, FF 21, DB 7, NFO Papers, ISU; *NFO Reporter*, Nov. 1971; NFO statement, "Family Farms," Sept. 1978, FF 14, DB 7, NFO papers, ISU.

44. President Lowell E. Gose, Iowa State Farmers Union state convention speech, Sept. 1974, FF PL 1.1.3 Farmers Union 1973–74, DB 86A, Clark Papers, UI; Oren Lee Stately, copy of statement to Senate Small Business Committee, Dec. 1, 1971, DB A0056B food hearings, Abourezk Papers, USD.

45. *Des Moines Register*, Nov. 24, 1971; Neil E. Harl, "Influencing the Structure of Agriculture," FF Corporate Farming, DB Gilbert Agricultural Files 200, Culver Papers, UI; William H. Scofield, "Agricultural Corporations Today," in *Corporate Farming and the Family Farm* (Ames: Iowa State University Press, 1970), 12–18; Introduction, "Impact of Corporation Farming on Small Business," Report of the Senate Select Committee on Small Business, Report no. 91-628, 3–4.

46. *Washington Post*, Apr. 25, 1972; McGovern, "New Light on the Family Farm," *Congressional Record*, 93d Cong., 1st sess, 119, pt. 82: 17616–25. He also inserted the speech by Rodefeld into the record. *Feedstuffs*, Apr. 2, 1973; Obey, "Threat of Corporate Farming," *Congressional Record*, 93d Cong., 2d sess., 120, pt. 3: 3512.

47. William J. Kuhfuss to Emanuel Celler, Mar. 21, 1972, John W. Scott, statement on corporate farming, Apr. 14, 1972, and Richard Rodefeld statement, Mar. 22, 1972, all in DB A0056C, Abourezk Papers, USD; Lowell E. Gose, speech to Iowa State FU convention, Sept. 1974, FF PL 1.1.3 Farmers Union 1973–74, DB 86A, MSC 414, Clark Papers, UI; Orville Freeman to Tony Dechant, Sept. 5, 1968, FF 47, DB 5, ser. 3, NFU Papers, UCB.

48. "In Pursuit of a Structures Policy: The Issue and the Role of USDA," Bergland speech to National Association of State Departments of Agriculture, Mar. 11, 1980, all in 151.H.11.6(F), DB 2, Bergland Papers, MHS; NFO *Reporter*, Apr. 1979; Fite, *American Farmers*, 220; Breimyer, *Over-fulfilled Expectations*, 242–44; *Star Tribune*, Dec. 30, 1987.

49. Virginia Grace Cook, "Corporate Farming and the Family Farm," Council of State Governments Research Brief, July 1976, 9; Neil E. Harl, "Farm Corporations: Present and Proposed Restrictive Legislation," Journal Paper J-6253, Iowa Agriculture and Home Economics Experiment Station, presented at North Central Workshop, "Corporation Farming: What Are the Issues?" Apr. 22, 1969; Curt Sorteberg to Kansas Farmers Union Land Policy Task Force Members, July 23, 1974, FF 30, DB 22, NFU Papers, UCB.

50. James Abourezk, "Agriculture, Antitrust, and Agribusiness: A Proposal for Federal Action," *South Dakota Law Review* 20 (summer 1975): 499–513.

51. David Weiman to Curt Sorteberg, Sept. 17, 1975, FF 30, DB 22, ser. 3, NFU Papers, UCB; Fred Shannon, *The Farmers' Last Frontier: Agriculture, 1860–1897* (New York: York, Rinehart & Winston, 1966), 154; Fite, *American Farmers*, 196; F. H. Buttel, "Agricultural Land Reform in America," in Geisler and Popper, *Land Reform, American Style*, 63.

52. Angus McDonald, "The Family Farm Is the Most Efficient Unit of Agricultural Production," Dec. 29, 1967, FF 17, DB 3, ser. 3, NFU Papers, UCB; *Washington Post*, Oct. 3, 1971.

53. Fite, *American Farmers*, 118; Willis E. Anthony, "The Corporate Form of Business," May 1970, Staff Paper P70-7, Department of Agricultural Economics, University of Minnesota, 17, 19; Kenneth R. Krause, *Corporate Farming: Importance, Incentives, and State Restrictions*, Dec. 1983, Economic Research Service, USDA, Agricultural Economic Report no. 506, ii; Jane Smiley, *A Thousand Acres* (New York: Fawcett Columbine, 1991), 4.

54. Unger, *Leaving the Land*, 180, 192, 194.

3. THE POLITICAL ECONOMY

1. Connor, Rogers, Marion, and Mueller, *Food Manufacturing Industries*, 9. The "first comprehensive examination of the organization of the food manufactur-

ing industries" was A. C. Hoffman's dissertation, which became the TNEC report *Large-Scale Organization in the Food Industries* (1940); Lee R. Martin, ed., *A Survey of Agricultural Economics Literature*, Economics of Welfare, Rural Development, and Natural Resources in Agriculture, 1940s to 1970s, vol. 3 (Minneapolis: University of Minnesota Press, 1981), 518; Willard F. Mueller, "Empirical Measurement in Market Structure Research," *Journal of Farm Economics* 43 (Dec. 1961): 1369; Willard F. Mueller and Robert Clodius, "Market Structure Research as an Orientation for Research in Agricultural Economics," *Journal of Farm Economics* 43 (Aug. 1961): 515–33.

2. Larry D. Umlauf (President, Pet Inc.) to Richard Clark, July 6, 1976, MSC 414, FF Ag 28 Food Marketing Commission, DB 47, Clark Papers, UI; Lewis Engman to Gerald Ford, Oct. 14, 1974, FF FTC, DB 5, Richard Cheney Papers, Ford Library; James Patton to the NFU Executive Committee, May 2 and 3, 1964, FF 25, DB 7, ser. 4, NFU Papers, UCB; Paul D. Scanlon, "FTC and Phase II: The McGovern Papers," *Antitrust Law and Economics Review* 5 (spring 1972): 19–36. The food processing industries have not received much attention from historians. Mark Wilde, "Industrialization of Food Processing in the United States, 1860–1960" (Ph.D. diss., University of Delaware, 1988), 1–2.

3. Donald Dewey, "Information, Entry, and Welfare: The Case for Collusion," *American Economic Review* 69, no. 4 (Sept. 1979): 587. Neil Fligstein notes that the focus of antitrust efforts from 1950 to 1974 was the level of concentration. In the 1960s more and more measures such as prices and profits were developed to measure market power. The change in antitrust philosophy in the early 1970s contributed to the fourth merger wave, which started in 1976 and continues to this day. Neil Fligstein, *The Transformation of Corporate Control* (Cambridge MA: Harvard University Press, 1990), 194. The FTC's National Food Commission of the 1960s did find the food industry "generally efficient and competitive," as had TNEC. G. E. Brandow, "The National Food Commission: Its Product and Role," *Journal of Farm Economics* 48 (Dec. 1966): 1322; Temporary National Economic Committee, Monograph no. 23, "Agriculture and the National Economy," 22.

4. A. C. Hoffman in Bruce Marion, ed., *The Organization and Performance of the U.S. Food System* (Lexington MA: Heath, 1986), xix–xxv; Bruce W. Marion, "Government Regulation of Competition in the Food Industry," *American Journal of Agricultural Economics* 61 (Feb. 1979): 181.

5. John M. Connor and Robert L. Wills, "Marketing and Market Structure of the U.S. Food Processing Industries," in Chester O. McCorkle Jr., ed., *Economics of Food Processing in the United States* (San Diego: Academic Press, 1988), 135; Tony Freyer, *Regulating Big Business: Antitrust in Great Britain and America, 1880–1990* (New York: Cambridge University Press, 1992), 270; Jon Didrichsen,

"The Development of Diversified and Conglomerate Firms in the United States, 1920–1970," *Business History Review* 46, no. 2 (summer 1972): 209–10. Didrichsen notes four general ways firms have grown since the 1920s: internal development driven by intrafirm research and supplemented by a few acquisitions of firms in closely related fields; acquisition of firms with related technologies, as opposed to internal development; acquisition of firms who use the same marketing and sales techniques (a good example of this trend is tobacco companies moving into food and alcohol sales); and the conglomerate strategy of acquiring firms in totally unrelated areas (a good example of this is Beatrice Foods moving into education and recreation). The decision to pursue a particular strategy hinges on the strengths of the firm. Didrichsen concludes that "even firms which acquire unrelated businesses frequently do have a body of knowledge or skill in general management or finance which they draw on in achieving their diversity," 219.

6. Dennis C. Mueller, "The Effects of Conglomerate Mergers: A Survey of the Empirical Evidence," *Journal of Banking and Finance* 1 (1977): 337; Mira Wilkins, *The Maturing of Multinational Enterprise: American Business Abroad from 1914 to 1979* (Cambridge MA: Harvard University Press, 1974), 329; William D. Heffernan and Douglas H. Constance, "Transnational Corporations and the Globalization of the Food System," in Alessandro Bonnano, Lawrence Busch, William H. Friedland, Lourdes Gouveia, and Enzo Mingione, eds., *From Columbus to ConAgra: The Globalization of Agriculture and Food* (Lawrence: University Press of Kansas, 1994), 29–49.

7. Willard Mueller and Thomas Paterson, "Policies to Promote Competition," in Marion, ed., *Organization and Performance of the U.S. Food System*, 387; Marion, "Government Regulation of Competition in the Food Industry," 180; James T. Halverson (Bureau of Competition Director, FTC), statement, June 28, 1973, DB 28, FF 13, Agribusiness Accountability Project Papers, ISU.

8. Mueller and Paterson, "Policies to Promote Competition," 384–85; Connor et al., *Food Manufacturing Industries*, 265–66. One-half of the complaints during this period involved milk, retailing, and beer. Very few complaints targeted grain belt commodity processors such as meatpackers, wheat millers, or corn and bean processors. The DOJ and the FTC brought 290 food system cases from 1956 to 1960, but only 64 from 1976 to 1980. John M. Connor and Robert L. Wills, "Marketing and Market Structure of the U.S. Food Processing Industries," in McCorkle, ed., *Economics of Food Processing*, 136.

9. Greig cites a study of the ninety-seven economies of scale studies conducted between 1950 and 1972 indicating that many food processing plants were below the "minimum efficient size." The studies indicated that in order to fully exploit economies of scale the total number of processing plants would need to be reduced

50 to 75 percent. Greig also notes that the building of an optimally sized processing plant is relatively inexpensive—relative to petrochemical complexes, steel mills, or automobile plants. Some economies also accrue in marketing, advertising, research, procurement, finance, and management. W. Smith Greig, "The Changing Structure of the Food Processing Industry: Description, Causes, Impacts and Policy Alternatives," Bulletin 827, Sept. 1976, College of Agriculture Research Center, Washington State University, 15–17; William G. Sheperd, "Causes of Increased Competition in the U.S. Economy, 1939–1980," *Review of Economics and Statistics* 64, no. 4 (Nov. 1982): 624. Some even doubt that competition was low in the 1950s. Clair Wilcox noted that "thirty cents of the consumer's dollar goes for food. Most of his crackers come from two concerns, but he can choose among several brands of bread or buy flour produced by any one of 2,000 packers, but pork comes from more than 500, beef from more than 600, and sausage from more than 1,100, and all of them must compete with poultry and fish." Clair Wilcox, "On the Alleged Ubiquity of Oligopoly," *American Economic Review* 40, no. 2 (May 1950): 70; Mueller, "Effects of Conglomerate Mergers," 344.

10. This is an early form of the strategy of raising rivals' costs. In the 1970s when states attempted their own inspections to protect state industries, similar to "protections" advanced in the 1890s, the multistate packers vehemently opposed it. Hormel, Armour, and Wilson spent four years in court seeking a ruling that federal regulations trumped state regulations. Byron M. Crippin Jr. (attorney for Hormel) to Richard Clark, June 20, 1973, MSC 414, FF Ag 4-1 Beef Processing-Feeding, DB 17, Clark Papers, UI. The vice-president of the American Meat Institute argued before Congress that the "situation becomes clearer when the Michigan standards are viewed for what they actually are—trade barriers that have the effect of making it difficult for a Federally inspected establishment endeavoring to produce for a national market to do business in the state." Alex P. Davies (Vice-President, American Meat Institute) to Subcommittee on Agricultural Research, Senate Agriculture Committee, June 14, 1973, speech in MSC 414, FF .G 4-1 Beef Processing-Feeding, DB 17, Clark Papers, UI.

11. Gary D. Libecap, "The Rise of the Chicago Packers and the Origins of Meat Inspection and Antitrust," *Economic Inquiry* 30 (Apr. 1992): 244, 246, 248, 258–59. Libecap doubts the arguments of farmers and smaller packers and simply sees them as victims of structural change in a competitive economy, casting doubt on the notion that the intent of the Sherman Act was competition and efficiency. See Thomas W. Hazlett, "The Legislative History of the Sherman Act Re-Examined," *Economic Inquiry* 30 (Apr. 1992): 263–76.

12. Robert M. Aduddell and Louis P. Cain, "Public Policy toward 'The Greatest Trust in the World,'" *Business History Review* 55, no. 2 (summer 1981): 220, 238–

40. Some historians, such as Alfred Chandler, see the emergence of the "big four" meatpackers resulting from an "inherent" advantage conferred by the nature of the industry. If true, the advantages of the arrangements quickly dissipated after the war, and the meatpacking industry was soon riven by crises that created a new set of competitive problems. As Mary Yeager sees it, "the high capital investments required to build and operate large, integrated firms reduced profit margins and created severe pressures to keep resources fully employed. These pressures, in turn, intensified competition" (*Competition and Regulation: The Development of Oligopoly in the Meat Packing Industry* [Greenwich CT: JAI Press, 1981], 242). In 1948 the DOJ began an effort to break Armour and Swift into five companies each and Cudahy and Wilson into two each. The effort was abandoned in 1954. In 1956 a federal judge refused to modify the 1920 consent decree. Jimmy M. Skaggs, *Prime Cut: Livestock Raising and Meatpacking in the United States, 1607–1983*, (College Station: Texas A&M University Press, 1986), 187–88.

13. C. Edward Harshbarger and Sheldon W. Stahl, "Economic Concentration in Agriculture: Trends and Developments," *Monthly Review*, Federal Reserve Bank of Kansas City (Apr. 1974), 23; Currier J. Holman to Richard Clark, Dec. 12, 1972, FF Ag 4 Meat Processing & Inspection, DB 17, Clark Papers, UI; Elmer C. Hunter, "Changes in the Cattle-Feeding Industry along the North and South Platte Rivers, 1953–1959," ERS 98, Farm Production Economics Division, USDA, in cooperation with the Colorado Agricultural Experiment Station, Mar. 1963 (Washington DC), iv; Willard Cochrane to Orville Freeman, Mar. 20, 1964, FF 1964 W. W. Cochrane memoranda (1), DB 11, 144.K.8.5, Freeman Papers, MHS; Currier J. Holman to Richard Clark, Dec. 12, 1972, DB 17, FF Ag 4 Meat Processing & Inspection, Clark Papers, UI. Ninety percent of cattle are fed on the large lots (one thousand or more) in the West, but only 15 percent are in Ohio, Indiana, Illinois, Wisconsin, and Minnesota. Iowa and Nebraska are in the middle. Gwen Quail, Bruce Marion, Frederick Geithman, and Jeffrey Marquardt, "The Impact of Packer Buyer Concentration on Live Cattle Prices," WP-89, May 1986, Working Paper Series, N. C. Project 117, 3; Teresa Glover and Leland Southard, "Cattle Industry Continues Restructuring," *Agricultural Outlook* (Dec. 1995), ERS, 14.

14. "Food from Farmer to Consumer," Report of the National Commission on Food Marketing, June 1966 (Washington DC), 22; Hunter, "Changes in the Cattle-Feeding Industry," 7; Samuel H. Logan, Lisa J. Steinmann, and Donald E. Farris, "Economics of Meat Processing in the United States," in McCorkle, ed., *Economics of Food Processing*, 247–48; Bruce Marion in Marion, ed., *Organization and Performance of the U.S. Food System*, 126. A study of fifteen western states in the 1960s indicated that most of the feedlots were owned by independent feeders, ranchers, and farmers and that packers and marketing firms controlled only about

12 percent of feedlot capacity in lots of one thousand or more head ("Food from Farmer to Consumer," 24). The Packers and Stockyards Administration ruled in the 1970s that packers couldn't own custom feedlots.

15. Aduddell and Cain, "Public Policy," 231–32; Robert M. Aduddell and Louis P. Cain, "The Consent Decree in the Meatpacking Industry, 1920–1956," *Business History Review* 55, no. 3 (autumn 1981): 364; John T. Schlebecker, *Cattle Raising on the Plains, 1900–1961* (Lincoln: University of Nebraska Press, 1963), 224; Homer Ayres to James Abourezk, hearing on beef imports, Sturgis SD, Apr. 14, 1977, FF 1, DB 2, ser. 2, NFU Papers, UCB; Yeager, *Competition and Regulation*, 240.

16. Willard Williams, "Small Business Problems in the Marketing of Meat and Other Commodities, Part 4: Changing Structure of Beef Packing Industry," Hearings before the Subcommittee on SBA and SBIC Authority and General Small Business Problems of the Committee on Small Business, House, 96th Cong., 1st sess. (Washington DC: GPO, 1979), 3 [hereinafter House Hearings]; "Food from Farmer to Consumer," 26; D. I. Padberg, "Economic Theory of Bargaining in Agriculture," *Journal of Farm Economics* 45, no. 5 (Dec. 1963): 1281; Roger Horowitz, "Meatpacking as Paradigm? Labor and the Dynamics of Industrial Change in 20th Century America," Research Seminar no. 15, Feb. 1994, Center for the History of Business, Technology, and Society, Hagley Museum and Library, 14; Connor et al., *Food Manufacturing Industries*, 104. For an example of packer competition for grocery store business (in this case Safeway), see J. F. Lambert to H. E. Williams, FF 2, DB 1, Rath Packing Company Committees (E), Rath Papers, ISU.

17. Speech by James Abourezk to Consumer Assembly, Jan. 30, 1975, DB 11, FF General 12 1973–75, Abourezk Papers, USD; Greig, "Changing Structure of the Food Processing Industry," 13–14; R. B. Heflebower, "Mass Distribution: A Phase of Bilateral Oligopoly or of Competition," Papers and Proceedings of the Sixty-Ninth Annual Meeting of the American Economics Association, Dec. 1956.

18. Schlebecker, *Cattle Raising*, 204, 224; Richard T. Crowder, "The Economic Outlook for Beef Cattle," 16, FF Soviet Grain Sales: Beef and Pork Prices, DB 96, Council of Economic Advisors Records, Ford Library; Bruce Marion, "A Comparison of Agricultural Subsectors," in Marion, ed., *Organization and Performance of the U.S. Food System*, 124.

19. E. W. Kuhn to B. J. Malusky, Mar. 20, 1970, FF FUGTA, DB 1, PI838, Kuhn Papers, MHS; Carl Curtis to Miles Kirkpatrick, Apr. 13, 1972, short synopsis, n.d., and Don to Office, Apr. 15, 1972, all in FF Federal Trade Commission, DB 80, Carl Curtis Papers, NSHS; "IBP and the U.S. Meat Industry," Harvard Business School case 9-391-006, revised Apr. 4, 1995, 3; Skaggs, *Prime Cut*, 168.

20. Wayne D. Purcell, "Economics of Consolidation in the Beef Sector: Research

Challenges," *American Journal of Agricultural Economics* 72 (Dec. 1990): 1212, 1214; John Connor, Dale Heien, Jean Kinsey, and Robert Wills, "Economic Forces Shaping the Food-Processing Industry," *American Journal of Agricultural Economics* 67 (Dec. 1985): 1137. Before the coming of the New Industrial Economics in the late 1970s price elasticity of demand as an "explanatory factor [was] largely ignored in previous structure-performance studies." John M. Connor, "Empirical Challenges in Analyzing Market Performance in the U.S. Food System," *American Journal of Agricultural Economics* 72 (Dec. 1990): 1224; *Independent Stockgrowers of America Newsletter*, c. 1974, FF 1, DB 2, ser. 2, NFU Papers, UCB; Skaggs, *Prime Cut*, 184; Harshberger and Stahl, "Economic Concentration in Agriculture," 21; Fite, *American Farmers*, 140; *Wall Street Journal*, Oct. 15, 1973.

21. "Beef and veal imports, which amounted to 1,037 million pounds in 1961, rose to 1,677 million pounds in 1963. The rise in imports coincided with increased domestic beef and veal production, which rose from 15,890 million pounds in 1961 to 16,896 million pounds in 1963. Although the imports were not of the same quality as most domestic beef and veal, their increase by 640 million pounds, added to the 1,006-million-pound rise in domestic veal and beef production, further depressed cattle prices in 1963" (*Report of the National Commission on Food Marketing*, 28).

22. Schlebecker, *Cattle Raising*, 220–21; Walter Hasty, "Recent Legislative Developments, Beef Imports in the Cattle Industry," Supplement 1, Legislative Analysis Memorandum 56-4, Feb. 10, 1964, NFU–NCFC Papers, ISU; J. W. Freebairn and Gordon C. Rausser, "Effects of Changes in the Level of U.S. Beef Imports," *American Journal of Agricultural Economics* 57 (Nov. 1975): 687–88.

23. L. William Seidman to Gerald Ford, Dec. 12, 1975, FF memo to President, 1975–May 1976, DB 76, William Seidman Papers, Ford Library; Wray Finney (American National Cattlemen's Association) to Gerald Ford, Aug. 27, 1976, and Henry Kissinger to Wray Finney, Oct. 5, 1976, FF Agricultural Policy Committee, DB 37, Seidman Papers, Ford Library; Clifford P. Hansen to Donald Rumsfeld, Feb. 18, 1972, FF Meat Imports, DB H, Butz Papers, PU; Gary Clyde Hufbauer, Diane T. Berliner, and Kimberly Ann Elliott, *Trade Protection in the United States* (Washington DC: Institute for International Economics, 1986), 323, 325; Bruce Marion in Marion, ed., *Organization and Performance of the U.S. Food System*, 130; Theodore White, *The Making of the President, 1972* (New York: Atheneum, 1973), 65.

24. Joe E. Owens to Senator Richard Clark, Sept. 1975, MSC 414, DB 38, FF Ag dept programs other than rural devel., Clark Papers, UI; Logan, Steinmann, and Farris, "Economics of Meat Processing," 252–53; "IBP and the U.S. Meat Industry," 6; Teresa Glover and Leland Southard, "Cattle Industry Continues Restruc-

turing," *Agricultural Outlook* (Dec. 1995), ERS, 15; *Value Line Investment Survey*, Feb. 16, 1996, part 3, *Ratings and Reports*, 10th ed., 1480. The U.S. signed free trade agreements in the late 1980s and early 1990s with the countries to which it exports the most agricultural products—Japan, Canada, and Mexico. With the growth of international trade even a disciple of the monopoly creed like John Kenneth Galbraith can acknowledge the competitiveness of the economy (*The Good Society: The Humane Agenda* [Boston: Houghton Mifflin, 1996], 16).

25. *Post-Bulletin* (Rochester MN), Mar. 24, 1966; *Ottumwa (IA) Courier*, May 5, 1971; *Farm Journal* 101, no. 1 (Jan. 1977): 37; statement of Walt Hackney, July 10, 1979, FF 17, DB 7, NFO Papers, ISU; *Chicago Sun-Times*, Sept. 7, 1962; D. W. Wright to Wm. M. Cameron, Aug. 14, 1967, D. W. Wright memo, Aug. 31, 1967, and Bernard W. Ebbing to Gene Potter, May 1, 1968, all in FF 9, DB 6, ser. F, Rath Papers, ISU; Preliminary Prospectus, Jan. 29, 1964, FF 12, DB 97, Financial T267, Rath Papers, ISU.

26. Mark Silbergeld (attorney, Consumers Union) speech, Dec. 12, 1973, DB SenA0056A, Abourezk Papers, USD; *Washington Post*, Nov. 8, 1971; Ray A. Goldberg, "Marketing Costs and Margins: Current Use in Agribusiness Market-Structure Analysis," *Journal of Farm Economics* 47 (Dec. 1965): 1356; Robinson is quoted in Willard F. Mueller, "Market Power and Its Control in the Food System," *American Journal of Agricultural Economics* 65 (Dec. 1983): 859.

27. Goldberg, "Marketing Costs and Margins," 1356–59; Brandow, "National Food Commission," 1322. The commission did recommend that firms be forced to report profits for specific fields, not just total company profits (p. 1324). In fact, the commission was created in response to the fears of cattlemen about the spread between farm and retail prices. Robert O. Aders, "The Food Industry and the Food Commission Report," *Journal of Farm Economics* 48 (Dec. 1966): 1328; Charles E. French, "Universities and the Food Commission Report," *Journal of Farm Economics* 48 (Dec. 1966): 1336. On the problem of policing a collusive agreement see George J. Stigler, "A Theory of Oligopoly," *Journal of Political Economy* 72 (Feb. 1964): 48–56.

28. Logan, Steinmann, and Farris, "Economics of Meat Processing," 250–51; Dale Tinstman and Robert Peterson, *Iowa Beef Processors, Inc.: An Entire Industry Revolutionized!* (New York: Newcomen Society, 1981).

29. Horowitz, "Meatpacking as Paradigm?" 20; Skaggs, *Prime Cut*, 197. Bruce Fehn, "'Chickens Come Home to Roost': Industrial Reorganization, Seniority, and Gender Conflict in the United Packinghouse Workers of America, 1956–1966," *Labor History* 34 (spring/summer 1993): 341. LTV was certainly part of the conglomerate wave: it also acquired Braniff International and Jones and Laughlin Steel.

30. "George A. Hormel & Company," Harvard Business School, May 7, 1991, case 9-591-026, 1, 4. The data on plant closings in the paragraph was taken from House Hearings part 3, *Beef in America: An Industry in Crisis*, 35–42, 45–46; Wilson J. Warren, "When 'Ottumwa Went to the Dogs': The Erosion of Morrell-Ottumwa's Militant Unionism, 1954–1973," *Annals of Iowa* 54, no. 3 (summer 1995): 240–44. AMK was a producer of bottle caps with sales of $40 million in 1967 but was able to buy Morrells, with sales of $800 million. Knowlton quoted in "George A. Hormel & Company," 3.

31. Hughes Bagley, House Hearings, part 5, *Anticompetitive Practices in the Meat Industry*, 9, 22; Willard Williams citing a National Economic Research Associates Inc. report, House Hearings, part 4, *Changing Structure of Beef Packing Industry*, 32–33; Skaggs, *Prime Cut*, 195. IBP did not start breaking down carcasses into boxed beef until they bought the Dakota City plant in 1967. Tinstman and Peterson, *Iowa Beef Processors*, 9.

32. *National Provisioner* 161, no. 1 (July 5, 1969); J. F. Lambert to H. E. Williams, Apr. 4, 1973, FF 2, DB 1, Rath Packing Company Committees (E), Rath Papers, ISU; W. L. Heubaum to C. J. Holman, J. R. Kemp, J. F. Haigler, P. F. Engler, S. Feldman, W. K. Holman, J. F. Roeser, W. E. Burns, J. W. Mueller, H. A. Bagley, R. M. Collins, D. A. Hartstack, Apr. 27, 1973, 1–2, Hughes A. Bagley to W. L. Heubaum, Aug. 2, 1972, internal IBP memo, 1, Don Hartstack to Stanley Feldman, Oct. 10, 1972, internal IBP memo, 5, all in House Hearings, part 5: *Anticompetitive Practices in the Meat Industry*, 50, 125, 140, 144–45.

33. Perry Haines to formula pricing committee, Dec. 8, 1975, House Hearings, part 5, *Anticompetitive Practices in the Meat Industry*, 11; Open Letter to the Livestock and Meat Industry, Western States Meat Packers Association Inc., May 21, 1969, FF 25, DB 9, Rath Packing Company Correspondence (F), Rath Papers, ISU. "Under present arrangements, conglomerate and integrated firms publish financial data only for their total activities and thus disclose no information about operations in particular fields such as their specialized competitors regularly publish" (*National Commission on Food Marketing*, 106).

34. Hughes Bagley, House Hearings, part 5, *Anticompetitive Practices in the Meat Industry*, 5–6, 9; Perry Haines (Vice-President of scheduling at IBP) to Roy Zider (Executive Vice-President), Oct. 14, 1975, internal IBP memorandum quoted by John M. Fitzgibbons, Special Counsel to the subcommittee: "The mechanism to achieve it is the establishment of outside carcass suppliers. The epitome of this type source would be the small independent packer who has laid off his luggers and is locked into our operation via the tram. . . . This direction of buying in more and selling out more will increase IBP's participation in the market, and therefore market share and control" (House Hearings, part 5, *Anticompetitive Practices in the Meat Industry*, 11.

35. Skaggs, *Prime Cut*, 198; House Hearings, part 3, *Beef in America: An Industry in Crisis*, 27–30, 144–47; Donald D. Stull and Michael J. Broadway, "The Effect of Restructuring on Beefpacking in Kansas," *Kansas Business Review* 14 (fall 1990): 13.

36. In the same case the Supreme Court also limited private suits to stop mergers. "Cargill, Inc. vs. Monfort of Colorado, Inc.," *Wisconsin Law Review* (1987), 507. In 1978 the top ten packers were IBP; Swift [Esmark]; MBPXL [soon Cargill]; Morrell [United Brands]; Armour [Greyhound]; Dubuque [privately held]; Spencer [bought in 1978 by Land O'Lakes]; Wilson [LTV—world's largest hog slaughterer; has closed all its beef plants]; National Beef Packing; Monfort (the info in parentheses isn't on the footnoted page); House Hearings, part 3, *Beef in America: An Industry in Crisis*, 34.

37. Warren, "When 'Ottumwa Went to the Dogs'," 234; Skaggs, *Prime Cut*, 189–90. Bruce Marion in Marion, ed., *Organization and Performance of the U.S. Food System*, 131; Richard J. Arnould, "Changing Patterns of Concentration in American Meat Packing, 1880–1963," *Business History Review* 45 (spring 1971): 26–27, 29; Danton, "The Decline of an Oligopoly," 45.

38. Willard Williams, House Hearings, part 4, *Changing Structure of Beef Packing Industry*, 69; *Report of the National Commission on Food Marketing*, 25; Yeager, *Competition and Regulation*, 239–41; Aduddell and Cain, "Consent Decree," 363. Schlebecker, *Cattle Raising*, 190, 206; Arnould, "Changing Patterns of Concentration," 29. Edwin R. O'Neill (formerly president and chairman of Western States Meat Packers Association), House Hearings, part 3, *Concentration Trends in the Meat Industry*, 43.

39. Skaggs, *Prime Cut*, 193; Michael J. Broadway and Donald D. Stull, "Rural Industrialization: The Example of Garden City, Kansas," *Kansas Business Review* 14 (summer 1991): 1; Wilson J. Warren, "The Heyday of the CIO in Iowa: Ottumwa's Meatpacking Workers, 1937–1954," *Annals of Iowa* 51 (spring 1992): 375, 383; Breimyer, *Over-fulfilled Expectations*, 247, notes part of the problem is "individual farmers, glorying in their independence." Arnould, "Changing Patterns of Concentration," 28; Horowitz, "Meatpacking as Paradigm?" 11.

40. Mark A. Grey, "Turning the Pork Industry Upside Down: Storm Lake's Hygrade Work Force and the Impact of the 1981 Plant Closure," *Annals of Iowa* 54 (summer 1995): 253; Logan, Steinmann, and Farris, "Economics of Meat Processing," 251; Tracy Rhato to Earl Butz, Apr. 23, 1973, MSC 414, DB 18, FF Ag 16-1 Livestock Prices, Clark Papers, UI; Oakley M. Ray (President, American Feed Manufacturers Association) to Richard Clark, July 20, 1976, MSC 414, DB 47, FF Ag 28 Food Marketing Commission, Clark Papers, UI.

41. Wilson J. Warren, "When 'Ottumwa Went to the Dogs'," 219; Shelton

Stromquist, *Solidarity and Survival: An Oral History of Iowa Labor in the Twentieth Century* (Iowa City: University of Iowa Press, 1993), 160; Roger Horowitz, "'This Community of Our Union': Shop Floor Power and Social Unionism in the Postwar UPWA," paper presented at the Center for Recent United States History Scholars' Seminar, Apr. 15, 1995, 5–8.

42. House Hearings, part 5, *Anticompetitive Practices in the Meat Industry*, 142; "IBP and the U.S. Meat Industry," 12–13; Grey, "Turning the Pork Industry Upside Down," 257–59; Donald D. Stull, "Cattle Cost Money: Beefpacking's Consequences for Workers and Communities," *High Plains Anthropologist* 14, no. 2 (fall 1994): 64.

43. Edwin R. O'Neill [formerly president and chairman of Western States Meat Packers Association], House Hearings, part 3, *Concentration Trends in the Meat Industry*, 43; "George A. Hormel & Company," 3. See also Hardy Green, *On Strike at Hormel: The Struggle for a Democratic Labor Movement* (Philadelphia: Temple University Press, 1990), and Dave Hage and Paul Klauda, *No Retreat, No Surrender: Labor's War at Hormel* (New York: Morrow, 1989).

44. Skaggs, *Prime Cut*, 198; Michael J. Broadway, "From City to Countryside: Recent Changes in the Structure and Location of the Meat-and Fish-Processing Industries," in Donald T. Stull, Michael J. Broadway, and David Griffith, eds., *Any Way You Cut It: Meat Processing and Small-Town America* (Lawrence: University of Kansas Press, 1995), 22.

45. Bruce W. Marion, "Government Regulation of Competition in the Food Industry," *American Journal of Agricultural Economics* 61 (Feb. 1979): 178, 180. In the 1980s, food processing companies "took advantage of the antitrust environment, [which was] more tolerant of big deals that it [had been] in decades, to increase their size and marketing clout"; "The New Food Giants: Merger Mania Is Shaking the Once-Cautious Industry," *Business Week*, Sept. 24, 1984, 133.

46. Bruce Marion and Donghwan Kim, "Concentration Change in Selected Food Manufacturing Industries: The Influence of Mergers vs. Internal Growth," *Agribusiness* 7 (1991): 415, 427, 429; Adesoji Adelaja, Rodolfo Nayga Jr., and Zafar Farooq, "Predicting Mergers and Acquisitions in the Food Industry," *Agribusiness* 15 (1999): 1–3.

47. Purcell, "Economics of Consolidation in the Beef Sector," 1214; Harold F. Breimeyer, House Hearings, part 1, *Meat Marketing*, 29–30.

48. Alvin Oliver to Milton Young, June 23, 1967, FF 20-246-11, Young Papers, UND; Paul L. Farris, Richard T. Crowder, Reynold P. Dahl, and Sarahelen Thompson, "Economics of Grain and Soybean Processing in the United States," in McCorkle, ed., *Economics of Food Processing in the United States*, 317; Marion, ed., *Organization and Performance of the U.S. Food System*, 158; Richard G.

Heifner, James L. Driscoll, John W. Helmuth, Mack N. Leath, Floyd F. Niernberger, Bruce H. Wright, "The U.S. Cash Grain Trade in 1974: Participants, Transactions, and Information Sources," Agricultural Economic Report no. 386, Sept. 1977, ERS, USDA (Washington DC), 3; Paul L. Farris, "The Grain Procurement Industry," in John R. Moore and Richard G. Walsh, eds., *Market Structure of the Agricultural Industries* (Ames: Iowa State University Press, 1966), 253.

49. Marion, ed., *Organization and Performance of the U.S. Food System*, 147, 149; Walter G. Heid, *Changing Grain Market Channels*, Marketing Economics Division, ERS-39, USDA, Nov. 1961 (Washington DC), ii, 2, 4.

50. ADM dossier, FF 6, DB 33, Far-Mar-Co Papers, KSU; Farris et al., "Economics of Grain and Soybean Processing," 343–46; *Kansas Farm News* (Oct. 1963) in FF 13, DB 2, Far-Mar-Co Papers, KSU; "Cost Components of Farm-Retail Price Spreads," Agricultural Economic Report no. 391, Nov. 1977, National Economic Analysis Division and the Commodity Economics Division, ERS, USDA (Washington DC), 24; Lehman B. Fletcher and Donald D. Kramer, "The Soybean Processing Industry," in Moore and Walsh, eds., *Market Structure of the Agricultural Industries*, 228.

51. Farris et al., "Economics of Grain and Soybean Processing," 127; John M. Connor and Robert L. Wills, "Marketing and Market Structure of the U.S. Food Processing Industries," in McCorkle, ed., *Economics of Food Processing in the United States*, 332–33.

52. *Cooperative Consumer*, May 15, 1965; Thomas J. Baerwald, *Minnesota Flour Milling*, Geography Department Background Paper no. 1, 1979, Science Museum of Minnesota, available from MHS; Michael W. Babcock, Gail L. Cramer, William A. Nelson, "The Impact of Transportation Rates on the Location of the Wheat Flour Milling Industry," *Agribusiness* 1, no. 1 (1985): 61–71; Farris et al., "Economics of Grain and Soybean Processing," 317; Marion, ed., *Organization and Performance of the U.S. Food System*, 321–26; Walter G. Heid, *Changing Grain Market Channels*, Marketing Economics Division, ERS-39, USDA, Nov. 1961 (Washington DC), 11; National Commission on Food Marketing, *Organization and Competition in the Milling and Baking Industries*, Technical Study no. 5, June 1966 (Washington DC), 8, 11, 14–15, 30, 45; D. Jerome Tweton, "The Business of Agriculture," in Clifford E. Clark, ed., *Minnesota in a Century of Change: The State and Its People since 1900* (St. Paul: Minnesota Historical Society Press, 1989), 266–67.

53. Heid, *Changing Grain Market Channels*, 11, 19.

54. See the example of how Phillip-Morris's acquisition of Miller Brewing Company provided Miller enormous resources for competition, in Connor et al., *Food Manufacturing Industries*, 251–53.

55. *National Macaroni Manufacturers Association v. Federal Trade Commission*, 345 F.2d 421, 422 (7th Cir., 1965).

56. Marion, ed., *Organization and Performance of the U.S. Food System*, 149; Farris et al., "Economics of Grain and Soybean Processing," 334, 336–38; National Commission on Food Marketing, *Organization and Competition in the Milling and Baking Industries*, 9.

4. THE GRAIN TRADING "CARTEL"

1. Unger, *Leaving the Land*, 153. The work cited most often is Dan Morgan, *The Merchants of Grain* (New York: Penguin, 1980), 14, 19, 33. What is not cited are the portions of the book that mention the "commercial rivalries of private grain companies," their "wide-open, disorganized trade," and the trade's "savagely competitive" nature (45, 58, 424); Clark statement, Multinational Corporations and U.S. Foreign Policy, International Grain Companies, Hearings before the Subcommittee on Multinational Corporations of the Committee on Foreign Relations, United States Senate, 94th Cong., June 18, 1976, part 16, 2 (hereinafter Senate Hearings); Food Action Campaign press release, Sept. 8, 1973, FF 34, DB 15, ser. 3, NFU Papers, UCB.

2. Senate Hearings, June 23, 1976, part 16, 61.

3. See generally Fite, *George N. Peek*; David E. Hamilton, *From New Day to New Deal: American Farm Policy from Hoover to Roosevelt, 1928–1933*, (Chapel Hill: University of North Carolina Press, 1991); Joan Hoff-Wilson, "Herbert Hoover's Agricultural Policies, 1921–1928," in Ellis W. Hawley, ed., *Herbert Hoover as Secretary of Commerce, 1921–1928: Studies in New Era Thought and Practice* (Iowa City: University of Iowa Press, 1981), 116–44.

4. George W. Stocking and Myron W. Watkins, *Cartels in Action: Case Studies in International Business Diplomacy* (New York: Twentieth Century Fund, 1946), 28, 39. The failure of the plan led to state intervention during the New Deal. The Sugar Act of 1937 established national quotas for the different sugar-producing regions in the United States. Also, just as in the case of wheat, an International Sugar Agreement was established in 1937 and an International Sugar Council created to manage the overproduction problem.

5. David Kilroy, "Extending the American Sphere to West Africa: Dollar Diplomacy in Liberia, 1908–1926" (Ph.D. diss., University of Iowa, 1995); Michael Hogan, *Informal Entente: The Private Structure of Cooperation in Anglo-American Economic Diplomacy, 1918–1928* (Columbia: University of Missouri Press, 1977), 207–8. The *St. Paul Pioneer Press* advocated using the demand for American wheat to wring concessions from European monopolists. Joseph Brandes, *Herbert Hoover and Economic Diplomacy: Department of Commerce Policy, 1921–*

1928 (Pittsburgh: University of Pittsburgh Press, 1962), 67–68, 84–105; Joseph Brandes, "Product Diplomacy: Herbert Hoover's Anti-Monopoly Campaign at Home and Abroad," in Hawley, ed., *Herbert Hoover*, 197, 200–201.

6. Hamilton, *From New Day to New Deal*, 110.

7. Peter B. Kenen, *Giant among Nations: Problems in United States Foreign Economic Policy* (Chicago: Rand McNally, 1960), 79; Leslie A. Wheeler, "Government Intervention in World Wheat Trade," *Journal of World Trade Law* 1 (July/Aug. 1967): 386; Lloyd C. Gardner, *Economic Aspects of New Deal Diplomacy* (Madison: University of Wisconsin Press, 1964), 24; Wilson Gee, *The Social Economics of Agriculture* (New York: Macmillan, 1942), 570. Gee argued that if "foreign markets are not regained, sweeping readjustments will have to be made in the entire farm economy of the nation," 18.

8. As Elaine Fuller sees it, "postwar debates over agricultural price supports recognized the importance of expanding agricultural exports but basically the issue was ignored. . . . What seemed to be emerging was a struggle between proponents of economic nationalism, represented by the system of farm price supports (whether rigid or flexible), income payments, tariffs and subsidies, in opposition to an internationalist position, the latter now represented by supporters of the State Department's strong position on creating a global multilateral free trade system" (Elaine Fuller, "American Wheat: From Surplus Production to Export Promotion, 1945–1975," paper presented at the Social Science History Association Conference, Nov. 1995, 17).

9. O. B. Jesness, "American Agriculture and Foreign Economic Policy," in *United States Agriculture: Perspectives and Prospects*, background papers and final report of Seventh American Assembly, Harriman NY, May 5–8, 1955, 85; Fuller, "American Wheat," 21, 23; Carmine Nappi, *Commodity Market Controls: A Historical Review* (Lexington MA: Heath, 1979), 37–47, 63; Colleen M. O'Connor, "Going against the Grain: The Regulation of the International Wheat Trade from 1933 to the 1980 Soviet Grain Embargo," *Boston College International and Comparative Law Journal* 5 (winter 1982): 239. The Hoover administration was under enormous pressure to agree to international quotas at an international conference held in May 1931 in London but refused to agree to the plans offered (Hamilton, *From New Day to New Deal*, 117–19).

10. Thomas W. Zeiler, *American Trade and Power in the 1960s* (New York: Columbia University Press, 1992), 23; Jane M. Porter and Douglas E. Bowers, *A Short History of U.S. Agricultural Trade Negotiations*, USDA, Economic Research Service, Agriculture and Rural Economy Division, Aug. 1989, 1–4; Charles P. Kindleberger, *Foreign Trade and the National Economy* (New Haven: Yale University Press, 1962), 133–34; Anne O. Krueger, *American Trade Policy: A Tragedy in the*

Making (Washington DC: American Enterprise Institute, 1995), 28; Edward S. Mason, *Controlling World Trade: Cartels and Commodity Agreements* (New York: McGraw Hill, 1946); Ronald T. Libby, *Protecting Markets: U.S. Policy and the World Grain Trade* (Ithaca NY: Cornell University Press, 1992), 21. The World Trade Organization was established in 1994 to do what the ITO was supposed to do. The WTO recently declared that American clean air laws violate trade rules, vindicating the fears of those of the left (about environmental and labor laws being undermined) and right (about sovereignty being undermined) and increasing the fears of those who believe that the U.S. will begin to substitute bilateral agreements and unilateralism for the rules of the WTO (Jeffrey E. Garten, "Is America Abandoning Multilateral Trade?" *Foreign Affairs* 74 [Nov.–Dec. 1995]: 50).

11. William F. Brooks (Executive Secretary, National Grain Trade Council) to Olin D. Johnston, FF International Wheat Agreement, 1949, DB 34, Agriculture Committee Files, Hickenlooper Papers, Hoover Library; B. F. Bowman to Bourke Hickenlooper, May 21, 1948, and V. B. Smith to Bourke Hickenlooper, May 20, 1948, FF Wheat Agreements, 1948, DB 31, Executive Department Files, Agriculture Department, Hickenlooper Papers, Hoover Library. Senator Hickenlooper was appointed to the subcommittee that studied the IWA bills. Tom Linder to Arthur Capper and Clifford R. Hope, Apr. 16, 1948, Henry E. Kuehn to Arthur Capper, May 13, 1948, Millers' National Federation Statement, May 15, 1948, William Brooks to Arthur Vandenberg, July 29, 1948, William Brooks to Henry Cabot Lodge, July 21, 1948, FF Agri Correspondence, all in DB 37, Capper Papers, KSHS; *The Gleaner*, May 1948, 9; Edward G. Cale and Oscar Zaglits, "Intergovernmental Agreements Approach to the Problem of Agricultural Subsidies," *Iowa Law Review* 34 (1948–49): 234–35; "World Importers Are Cool to New Wheat Agreement," *Journal of Commerce*, May 1, 1956; Ted Rice, "The Influence of Government Programs on the Originating, Storing, and Exporting of Grain," *Journal of Farm Economics* 45 (Dec. 1963): 1317; *Capper's Weekly*, May 29, 1948.

12. Clarence B. Randall to the Council on Foreign Economic Policy, June 28, 1958, and Douglas Dillon and Don Paarlberg to Clarence Randall, June 13, 1958, FF U.S. Policy with Respect to IWA, DB 6, Council on Foreign Economic Policy Papers, Eisenhower Library; Department of Agriculture Proposal on U.S. Participation in International Wheat Agreement, Oct. 5, 1955, and Felix Wormser to Joseph Dodge, Feb. 24, 1956, FF U.S. Participation in IWA, DB 5, U.S. Council on Foreign Economic Policy Paper, Policy Paper Series, Eisenhower Library; "Wheat Production: Trends—Problems—Programs—Opportunities for Adjustment," Agriculture Information Bulletin no. 179, Mar. 1958, Agricultural Research Service, USDA, 45, 47; O'Connor, "Going against the Grain," 240, 242. Since the government agreed to supply wheat at prices much lower than the domestic support price,

the U.S. Treasury had to make up the difference. Murray R. Benedict, *Can We Solve the Farm Problem? An Analysis of Federal Aid to Agriculture* (New York: Twentieth Century Fund, 1955), 317–18. It would seem that AAA production restrictions and IWA quotas would have been illegal under the charter of the ITO, but separate provisions were made for international commodity agreements. Still, the international agreements were legal only if a "burdensome surplus" existed or a "widespread unemployment or under-employment" would occur in their absence, and if they were constituted among governments, included consumers and producers, and were limited to five years. Stanley Metzger, "Cartels, Combines, Commodity Agreements, and International Law," *Texas International Law Journal* 11 (1976): 528, 535–36.

13. "International Wheat Agreement," FF Background Material: Soviet Grain Sales, DB E22, Arthur Burns Papers, Ford Library; O'Connor, "Going against the Grain," 248–53; Tony Dechant, statement to the Special Senate Foreign Relations Subcommittee on the IWA, June 29, 1971, FF 36, DB 11, ser. 3, NFU Papers, UCB; Malmgren and D. L. Schlechty, "Rationalizing World Agricultural Trade," *Journal of World Trade Law* 4 (July/Aug. 1970): 536; Liaquat Ali, "The World Wheat Market and International Agreements," *Journal of World Trade Law* 16 (Jan.–Feb. 1982): 66–67; draft of NFO statement to Congress, FF 7 DB 7, NFO Papers, ISU; NFO *Reporter*, June 1972; *National Farmers Union Washington Newsletter*, May 23, 1968; Martha Hamilton to Jerry Berman and Jim Hightower, Sept. 3, 1971, FF 6, DB 1, Agribusiness Accountability Project Papers, ISU; "A Statement of Principles and Policies on Farm and Rural Issues," National Rural Coalition, DB A0056A, Abourezk Papers, USD; Morgan, *Merchants of Grain*, 183, 464; D. Gale Johnson, *World Agriculture in Disarray* (London: Macmillan, 1973), 256–57; Dale Lyon to John Sparkman, June 25, 1971, FF 36, DB 11, ser. 3, NFU Papers, UCB.

14. The first quarter of 1949, when the Marshall Plan was fully operational, it financed 83 percent of American corn exports and 67 percent of wheat flour. Fuller, "American Wheat," 19.

15. Even in May 1954 Dulles was still wary. John Foster Dulles to Ezra Taft Benson, May 12, 1954, DB 4, FF Disposal of U.S. Surpluses Abroad, Francis Papers, Eisenhower Library; Trudy Peterson, *Agricultural Exports, Farm Income, and the Eisenhower Administration* (Lincoln: University of Nebraska Press, 1979), 25; Robert M. Macy to Robert Cutler, May 7, 1953, White House Central Files, DB 502, OF 106-G-1, FF 106-I Agricultural Surpluses, Eisenhower Library; *Congressional Record*, 85th Cong., 1st sess. (June 14, 1957), 103, pt. 7: 8244–45; Roland A. Ouellette, American Law Division, Legislative Reference Service, "Development of the Agricultural Trade and Development and Assistance Act of 1954 in the 83rd Congress," Apr. 24, 1959, 3–4; Memorandum of Conversation between Mil-

ton Young and D. A. Fitzgerald, Mar. 26, 1954, FF Reading File Jan. 1, 1954–Mar. 31, 1954 (1), DB 35 A74-6, Fitzgerald Papers, Eisenhower Library.

16. Walter Bedell Smith to Sherman Adams, Feb. 11, 1954, White House Central Files, DB 502, OF 106-G-1, FF 106-1 Agricultural Surpluses, Eisenhower Library; Trudy Peterson, *Agricultural Exports*, 18–42.

17. J. Stuart Russell, "Surpluses," *Christian Century* 72 (Mar. 16, 1955): 338; Jacob J. Kaplan, *The Challenge of Foreign Aid: Policies, Problems, and Possibilities* (New York: Praeger, 1967), 50, 52; Don Paarlberg, "Essentials of Modern Trade Policy," in R. J. Hildreth, ed., *Readings in Agricultural Policy* (Lincoln: University of Nebraska Press, 1968), 235–36; Wheeler, "Government Intervention in World Wheat Trade," 387; Irwin R. Hedges, "Building Export Markets for American Wheat," speech to the National Association of Wheat Growers, and Raymond A. Ioanes, "The Soybean Industry's Stake in the Export Market," speech to the American Soybean Producers Association, Aug. 21, 1962, FF Trade: Common Market, DB 6, 67A1881, McGovern Papers, MLPU; Bureau of the Budget memo to LBJ, Feb. 7, 1967, DB 270, White House Central File, and Joe Califano to LBJ, Oct. 18, 1965, Executive Series, LBJ Library.

18. Alex F. McCalla and Andrew Schmitz, "Grain Marketing Systems: The Case of the United States versus Canada," *American Journal of Agricultural Economics* 61 (May 1979): 204–9; Michael J. McGarry and Andrew Schmitz, *The World Grain Trade: Grain Marketing, Institutions, and Policies* (Boulder CO: Westview, 1992), 333–34.

19. Reuben Johnson and Walter Hasty, "The Kennedy Round: GATT," Legislative Analysis Memorandum 7-64, July 21, 1964, FF 5, DB 1, NFU–NCFC Papers, ISU; Willard Cochrane to Orville Freeman and Undersecretary of Agriculture, Nov. 30, 1962, DB 10, 144.K.8.4, Freeman Papers, MHS; Kenneth Naden to Christian Herter, Sept. 29, 1966, FF 10, DB 7, ser. 4, NFU Papers, UCB; Paul Findley, *The Federal Farm Fable* (New Rochelle NY: Arlington House, 1968), 189; Alex F. McCalla and Andrew Schmitz, "State Trading in Grain," in M. M. Kostecki, ed., *State Trading in International Markets: Theory and Practice of Industrialized and Developing Countries* (New York: St. Martin's Press, 1982), 66–67. Agriculture accounted for 21 percent of American exports. *U.S. Agriculture and the Balance of Payments, 1960–1967*, USDA, Economic Research Service, Foreign 224 (Washington DC, Apr. 1968), 3; Morgan, *Merchants of Grain*, 186. The membership of Britain in the EEC was subsequently vetoed by France. Prior to the organization of the CAP the U.S. simply dealt with European nations on a one-to-one basis. For example, the U.S. negotiated with West Germany on the same terms that France did.

20. The previous system only allowed the administration to proceed with tedious product-by-product negotiations. The Europeans, however, wanted to nego-

tiate using a "linear" approach that would include all the products in broad economic categories.

21. Zieler, *American Trade and Power*, 37, 59, 61–62, 65–67, 170. Some members of the Kennedy administration sought trade liberalization on the grounds that Western protectionism alienated the Third World and aided Soviet efforts to win client states. Some believed that the Food for Peace program had undermined American influence in the Third World because it limited markets for Third World commodity exports. Hence, another situation emerged in which the interests and priorities of the American political economy were at loggerheads with the interests and priorities of the foreign policy makers.

22. Willard W. Cochrane to Orville Freeman, Jan. 7, 1964, FF 1964 W. W. Cochrane memoranda (1), DB 11, 144.K.8.5(B), Freeman Papers, MHS; Malmgren and Schlechty, "Rationalizing World Agricultural Trade," 517; Findley, *Federal Farm Fable*, 194–95; Raymond Ioanes, "Recent Common Market Developments and U.S. Agriculture," speech to National Farm Institute, Feb. 15, 1963, and Orville Freeman to GTA, Dec. 11, 1962, FF Trade: Common Market, DB 6, 67A1881, McGovern Papers, MLPU; Johnson, *World Agriculture in Disarray*, 25, 254–56. The CAP variable import levies were calculated to raise the price of an imported commodity to the level of the EEC price, eliminating the price advantages of imports.

23. Ross B. Talbott, *The Chicken War: An International Trade Conflict between the United States and the European Economic Community, 1961–64* (Ames: Iowa State University Press, 1978), ix, 3–4, 12, 54, 56, 84, 96, 113, 115, 166; Orville Freeman, speech to the Institute of American Poultry Industries, Feb. 15, 1963, FF Trade: Common Market, DB 6, 67A1881, McGovern Papers, MLPU; Willard Cochrane to Orville Freeman, June 16, 1964, FF W. W. Cochrane memoranda (1), 144.K.8.5(B), Freeman Papers, MHS.

24. *Milling and Baking News*, Mar. 1, 1983, 7, 9, and Mar. 22, 1983, 67; *New York Times*, Oct. 19, 1982, and Aug. 8, 1983; Libby, *Protecting Markets*, 24–26, 32, 58–60, 66.

25. Earl Butz speech to the Ft. Wayne Press Club, Apr. 21, 1972, DB M, Binder 12-6-71–5-4-72, Earl Butz Papers, PU; "Mr. (Neal) Smith, Home from Washington," *Des Moines Register*, Feb. 20, 1996, excerpt from forthcoming Neal Smith, *Mr. Smith Went to Washington: From Eisenhower to Clinton* (Ames: Iowa State University Press, 1996); Jon Lauck, "Binding Assumptions: Karl E. Mundt and the Vietnam War, 1963–1969," *Mid-America* 76 (fall 1994): 294–96; *Congressional Quarterly Almanac*, 1963, 327; Robert L. Paarlberg, *Food Trade and Foreign Policy: India, the Soviet Union, and the United States* (Ithaca NY: Cornell University Press, 1985), 72; Robert Paarlberg, "Agriculture in the Eisenhower and Nixon

Years," Indiana University Oral History Research Project, Mar. 1978, 6, Special Collections Department, PU; George McGovern, U.S.-Soviet Wheat Sales Speech, Mar. 20, 1964, FF Ag. Farm Program, General, DB Senate File, 1968, McGovern Papers, MLPU; George McGovern to Ralph Massey, LBJ, George Meany, and Ray Murdock, Aug. 26, 1965, General File, LBJ Library. Fuller notes the "grain traders were the first to make a definitive breach in Cold War export controls policy" ("American Wheat," 3).

26. NCFC press release, June 24, 1971, FF 10, DB 1, NFU–NCFC Papers, ISU; Ali, "World Wheat Market," 60; George P. Schultz, statement on foreign economic policy, Feb. 12, 1973, FF Council on International Economic Policy, White House, 1973, DB G, and Earl Butz speech to NCFC, Jan. 12, 1972, DB M, binder 12-6-71–5-4-72, both in Butz Papers, PU; Robert Pastor, *Congress and the Politics of U.S. Foreign Economic Policy, 1929–1976* (Berkeley and Los Angeles: University of California Press, 1990), 129, 138. President Nixon was responding to international economic developments that precipitated the conversion of American dollars into gold in the late 1960s. In May 1971 four European countries revalued their money upward, effectively ending the postwar Bretton Woods international monetary system, based on the American dollar. In turn, "the president unilaterally suspended dollar convertibility into gold, effectively devaluing it, and instituted a wage freeze, a tax surcharge on imports, and a series of measures to improve U.S. export performance—with increased *agricultural* exports the center of this strategy" (James Wessel, *Trading the Future* [San Francisco: Institute for Food and Development Policy, 1983], 161); Morgan, *Merchants of Grain*, 15, 172, 193, 197, and 207; Herbert Stein, "Food Prices: Oh, How It All Adds Up," *New York Times*, Sept. 29, 1972; Speech by Senator McGovern in Vernon Center MN, Sept. 11, 1972, 7, FF McGovern, DB H, Butz Papers, PU; McGovern, "People's Dinner" speech, 62; Ken W. Clawson to David Gergen, Mar. 9, 1973, DB F, FF Agriculture, Butz Papers, PU. For internal NFU debate on the wisdom of a lawsuit against USDA and grain company officials for collusion see Weldon Barton to Tony Dechant, Nov. 1, 1972, and Medill Barnes to Jay Naman, Dec. 26, 1972, FF 25, DB 13, ser. 3, NFU Papers, UCB.

27. A. V. Krebs, *The Corporate Reapers: The Book of Agribusiness* (Washington DC: Essential Books, 1992), 303; Agribusiness Accountability Project, "Business as Usual: Corporate Influence in Food Policy," Dec. 1973, and James McHale, Dec. 1973, DB SenA0056A, Abourezk Papers, USD; Morgan, *Merchants of Grain*, 214; Harold E. Wills (Far-Mar-Co) to Dick Clark, May 9, 1973, FF Ag 12-1, DB 18, Clark Papers, UI; Corporate Secrecy: Agribusiness, Hearings before the Subcommittee on Monopoly, 92d Cong., 1st and 2d sess., 1973, part 3; *Center for Rural Affairs Newsletter* 1 (Oct. 1973), Center for Rural Affairs Papers, ISU; Seymour Hersh, *The Price of Power: Kissinger in the Nixon White House* (New York: Sum-

mit, 1983), 335; *National Farmers Union Newsletter*, May 23, 1968; NFO *Reporter* 16 (June 1972): 3; Tony Dechant speech to the Annual Stockholders Meeting of CENEX, Mar. 15, 1972, FF 43, DB 2, ser. 2, NFU Papers, UCB.

28. Frank Church to Gerald Ford, Oct. 1, 1975, DB 4, FF 12, Frank Church Papers, Boise State University; Roger B. Porter, *The U.S.-U.S.S.R. Grain Agreement* (Cambridge MA: Harvard University Press, 1984), 44, 67, 104; *Top-Producer* (mid-Mar. 1996), 12; Joan Hoff, *Nixon Reconsidered* (New York: Basic Books, 1994), 204. The importance of agriculture in U.S.–Russian relations is still strong. When Russia was threatening to ban imports of American poultry President Clinton stepped in: "This is a big issue," Clinton told Yeltsin, "especially since 40% of U.S. poultry is produced in Arkansas" (*Newsweek*, Apr. 8, 1996).

29. Kaplan, *Challenge of Foreign Aid*, 50; Hersh, *Price of Power*, 346; William T. Weber, "The Complexities of Agripower: A Review Essay," *Agricultural History* 52 (Oct. 1978): 526–28. See also Orville Freeman, "Malthus, Marx, and the North American Breadbasket," *Foreign Affairs* 45 (July 1967): 579–93.

30. Morgan, *Merchants of Grain*, 339, 342, 335. As early as 1972 Secretary Butz hinted at a grain-for-oil arrangement (Opening remarks, Russian Trade Conference, Apr. 11, 1972, DB M, binder 12-6-71–5-4-72, Butz Papers, PU); James L. Mitchell to Dick Cheney, Sept. 6, 1975, FF Grain Sales to Soviet Union 8/75–10/75, DB 5, Cheney Papers, Ford Library.

31. William Schneider, *Food, Foreign Policy, and Raw Materials Cartels* (New York: Crane, Russak, 1976), ix, 57; Emma Rothschild, "Food Politics," *Foreign Affairs* 54 (Jan. 1976): 285, 294; Paarlberg, *Food Trade and Foreign Policy*, 21–27.

32. Paul MacAvoy to Allen Greenspan, Aug. 22, 1975, and L. William Seidman to Gerald Ford, Sept. 8, 1975, FF Grain Sales to Soviet Union 8/75–10/75, DB 5, Cheney Papers, Ford Library; Weber, "Complexities of Agripower," 526–28; Clifton Cox (President, Armour) to Arthur Burns, July 6, 1973, FF Wheat Embargo, 1973, DB B115, Burns Papers, Ford Library; Porter, *U.S.-U.S.S.R. Grain Agreement*, 68; Richard Gilmore, "Grain in the Bank," *Foreign Policy* 38 (spring 1980): 168; "Staley Testifies Grain Embargoes Cost Farmers Millions," Jan. 23, 1976, FF 18, DB 7, NFO Papers, ISU; Speech by Gerald Ford to the Farm Forum, Springfield IL, Mar. 5, 1976, 1, DB L, FF White House Agricultural Policy Committee, 1976, Butz Papers, PU; Porter, *U.S.-U.S.S.R. Grain Agreement*, 100–101; *On the Move*, Jan. 22, 1976; FF Dole Statements: Export Sales, DB F28, President Ford Committee Records, 1975–76, Ford Library; Robert Ray to Gerald Ford, Oct. 7, 1974, TA 3/00, 158, White House Central Files, Ford Library; George Van Cleve to Jim Cavanaugh, Aug. 10, 1976, FF Farm Policy, DB 4, Gergen Papers, Ford Library; Remarks by Jimmy Carter at Iowa State Fairgrounds, Aug. 25, 1976, FF Jimmy Carter, Ag. Campaign, DB F28, Ford Committee Records, 1975–76, Ford Library;

Randy George Bush, "The Political Economy of Wheat Production and Exporting in the Pacific Northwest, c. 1848–1984" (Ph.D. diss., University of Washington, 1986), 354; Paarlberg, *Food Trade and Foreign Policy*, 11–12.

33. *Des Moines Register*, June 12, 1972; Henry Kissinger, *White House Years* (Boston: Little, Brown, 1979), 1270. Kissinger recalled that "our intelligence about Soviet needs was appalling." Food Deputies Group to Gerald Ford, Sept. 6, 1975, FF Soviet Grain Sales: Food Deputies Group, DB 96, Council of Economic Advisors Papers, Ford Library; Jackson press releases, July 12, 1975, DB 262, B497e and FF 138, DB 260, B497c, Henry Jackson Papers, University of Washington; Gene Moos to Earl Butz, Sept. 11, 1972, FF 36, DB 14, ser. 3, NFU Papers, UCB; Porter, *U.S.-U.S.S.R. Grain Agreement*, 13–17; Morgan, *Merchants of Grain*, 208–9, 269, 280; Myron Just, Senate Hearings, June 18, 23, 24, 1976, part 16, 38.

34. Morgan, *Merchants of Grain*, 29; Larry Ziegler (President of North Dakota NFO), Senate Hearings, June 18, 23, 24, 1976, part 16, 45; Wheeler, "Government Intervention in World Wheat Trade," 396; Malmgren and Schlechty, "Rationalizing World Agricultural Trade," 536; Ali, "The World Wheat Market and International Agreements," 59; O'Connor, "Going Against the Grain," 249, 259–61; letter from thirty-two senators to William Proxmire, June 26, 1980, FF Soviet Grain Embargo: 7-17-80, DB 74, M80-626, Nelson Papers, WHS; Emma Rothschild, "Food Politics," *Foreign Affairs* 54 (Jan. 1976): 290, 298; *Oil and Gas Journal*, June 18, 1979.

35. *NFO Reporter*, Dec. 1975; NFU press release, Oct. 4, 1972, FF 223, DB 3, ser. 2, NFU Papers, UCB; Morgan, *Merchants of Grain*, 120, 322–23; Bill Wagner, "The Widening Horizons of the GTA," *Corporate Report* (Sept. 1976): 38, 65; Senate Hearings, June 23, 1976, part 16, 38, 41–42; *Farm Journal*, Feb. 1977, 56; Marion, *Organization and Performance of the U.S. Food System*, 156; "Market Structure and Pricing Efficiency of U.S. Grain Export System," GAO/CED-82-61, June 15, 1982, 17; McGarry and Schmitz, *World Grain Trade*, 481.

36. Morgan, *Merchants of Grain*, 262–67, 432–40; "Notice to the Press," Oct. 5, 1974, TA 3/100 158, White House Central Files, Ford Library; Porter, *U.S.-U.S.S.R. Grain Agreement*, 27; *Wall Street Journal*, Apr. 8, 1977.

37. *Des Moines Register*, Nov. 12, 1973; *On the Move* (NFO), Feb. 23, 1973; *The NFO Family Farmer*, FF 24, DB 4, NFO Papers, ISU; Truman David Wood, "The NFO in Transition" (masters thesis, University of Iowa, 1961), 38; *NFO Reporter*, June 1972 and June 1976; George Brandsberg, *The Two Sides in NFO's Battle* (Ames: Iowa State University Press, 1964), 253.

38. *Farm Journal*, Feb. 1977, 56; *Key to the News* (Kansas City–St. Joseph), Dec. 24, 1972; Cook Industries v NFO, judgment ordered in favor of plaintiff by U.S. District Court for the Western District of Tennessee, Western Division, May

17, 1974, copy in FF 34, DB 8, NFO Papers, ISU; *Businessweek*, July 23, 1984; Ronald Knutson speech to the Farmers Grain Dealers Association of Iowa, Jan. 28, 1975, FF 22, DB 56, Far-Mar-Co Papers, KSU; Larry Ziegler, Senate Hearings, June 23, 1976, part 16, 45–46; "NFO's Memorandum for Meeting, Trade Agreement with Brazil, Feb. 1981," FF 32, DB 7, and Devon R. Woodland to Richard E. Lyng, Mar. 28, 1988, FF 3, DB 12, NFO Papers, ISU; *Kansas City Times*, Nov. 30, 1985; NFO *Reporter*, Apr. 1979; McGarry and Schmitz, *The World Grain Trade*, 488.

39. Walter B. Saunders, Senate Hearings, June 24, 1976, part 16, 101, 105, 110, 128–29; Wagner, "The Widening Horizons of the GTA," 38; Morgan, *Merchants of Grain*, 142; *Des Moines Register*, June 12, 1972.

5. THE NFO AND FARM BARGAINING

1. Fite, *American Farmers*, 102, 107, 116; James N. Giglio, *The Presidency of John F. Kennedy* (Lawrence: University Press of Kansas, 1991), 107; *Rapid City Journal*, June 19, 1960.

2. Daniel Nelson, *Farm and Factory: Workers in the Midwest, 1880–1990* (Bloomington: Indiana University Press, 1995), 170, 172.

3. William D. Anderson, "The Mission, History, and Times of the National Farmers Organization" (masters diss., University of Chicago, 1965), 52.

4. *Des Moines Register*, Oct. 21 and Dec. 16, 1955; Dan Turner to Mr. Wood, Jan., 5, 1961, FF NFO Correspondence 1955–61, DB 3, Turner Papers, UI; *Des Moines Register*, Aug. 8, 1957; Brandsberg, *Two Sides in NFO's Battle*, 66–67.

5. *Adams County (IA) Free Press*, Sept. 22, 1955; *Des Moines Register*, Dec. 16, 1955; minutes of NFO meeting, Jan. 9, 1956, FF 6, DB 13, NFO Papers, ISU; Resolutions-NFO, Adopted at National Convention, Nov., 1956, FF 5, DB 7, NFO Papers, ISU; Harold Breimyer to author, Nov. 18, 1996; "What's Behind the New Farm Crisis," *Business Week* (Dec. 10, 1955), 122; Fite, *American Farmers*, 158. In Fite's account, he implies that the NFO advocated collective bargaining from its inception in 1955. That approach was not adopted until the organization's annual convention in 1958. William C. Pratt, "The Farmers Union, McCarthyism, and the Demise of the Agrarian Left," *Historian* 58 (winter 1996): 333–34, 339–40; Truman David Wood, "The NFO in Transition" (Ph.D. diss., University of Iowa, 1961), 139; NFO *Reporter*, May 1956; Walters, *Holding Action*, 16.

6. TV-10 News, 1977, video cassette 15/25, NFO Papers, ISU; *St. Joseph News Press*, Nov. 19, 1957; Emery E. Jacobs to Tony T. Dechant, cc Charles Brannan, Oct. 11, 1957, 1957 Iowa NFO Convention resolutions, Emery E. Jacobs to Maurice O'Reilly and Leonard Hoffman, Emery E. Jacobs to Art Thompson, Jan. 17, 1958, all in FF 13, DB 7, ser. 4, NFU Papers, UCB; memo, Oren Lee Staley to G. A. Letterson, n.d., FF 8, DB 12, NFO Papers, ISU; *Des Moines Register*, Oct. 29,

1957. Reuther was a "strong supporter of the Farmer Labor parties that had sprung up in several Midwestern cities and states" (Nelson Lichtenstein, *The Most Dangerous Man in Detroit: Walter Reuther and the Fate of American Labor* [New York: Basic Books, 1995], 58–59).

7. *Adams County (IA) Free Press*, Sept. 22, 29, 1955; NFO pamphlet, FF 25, DB 4, NFO Papers, ISU; Brandsberg, *Two Sides of NFO's Battle*, 218; Anderson, "Mission, History, and Times," 10; Walters, *Holding Action*, 16; *Des Moines Register*, Oct. 29, 1957; "Here Are Facts About the NFO," FF 8, DB 7, NFO Papers, ISU; J. D. Bowser to Dan Turner, Aug. 4, 1957, and Dan Turner to Mr. Wood, Jan., 5, 1961, FF NFO Correspondence 1955–61, DB 3, Turner Papers, UI; *St. Joseph News-Press*, Nov. 19, 1957; NFO U.S. Farm Report, Jan. 15, 1968, film 166, NFO Papers, ISU; Wood, *NFO in Transition*, 19–21; Brandsberg, *Two Sides in NFO's Battle*, 74; *Des Moines Register*, Mar. 16, 1956; Oren Lee Staley, "Farmers Must Have Assistance to Help Themselves," statement to Senate Agriculture Committee, Jan. 16, 1978, FF 3, DB 12, NFO Papers, ISU. For the postwar influence of Reuther see, for example, Nelson Lichtenstein, "From Corporatism to Collective Bargaining: Organized Labor and the Eclipse of Social Democracy in the Postwar Era," in Steve Fraser and Gary Gerstle, eds., *The Rise and Fall of the New Deal Order, 1930–1980* (Princeton NJ: Princeton University Press, 1989), 126–27. The UAW was the "largest and most visible" of the CIO unions that boomed in the 1930s and 1940s. Nelson, *Farm and Factory*, 151–52. For earlier farm cooperation with the Farm Equipment Workers (FE) and the UPWA, see Wilson J. Warren, "The 'People's Century in Iowa: Coalition-Building among Farm and Labor Organizations, 1945–1950," *Annals of Iowa* 49, no. 5 (summer 1988): 376–82, 386–87.

8. NFO U.S. Farm Report, Jan. 15, 1968, film 166, NFO Papers, ISU; Kermit Veum, minutes of joint GTA. CENEX. NFU meeting, FF 26, DB 22, ser. 3, NFU Papers, UCB; *NFO Reporter*, Nov. 1974; TV-10 News, 1977, videocassette 15/25, NFO Papers, ISU; *Courier-Journal*, Dec. 14, 1973; "CED?—Certain Economic Destruction," *Farm Tempo USA*, Feb. 1970; Anderson, "Mission, History, and Times," 30; Fite, *American Farmers*, 146, 153, 166; Brandsberg, *Two Sides in NFO's Battle*, 90–97; *NFO Reporter*, June 1972; Wood County (WI) Chapter of NFO to Senate Subcommittee on Monopoly, July 22, 1968, FF Agriculture, Corporate Farming, DB 255, M74-549, Nelson Papers, WHS; *County Progressive*, Nov. 1970; U.S. Farm Report, "Shocking Planned Peasantry for the Independent Farmer," film 260, NFO Papers, ISU; Rowell, *Mad as Hell*, 26–28; *Economic Policy for American Agriculture*, Research and Policy Committee, Committee for Economic Development (Jan. 1956), 34; *Center for Rural Affairs Newsletter*, Aug. 1973, FF 1/4, DB T-179, Center for Rural Affairs Papers, ISU; reprint of "young executives" report, *Con-*

gressional Record, 92d Cong., 2d sess. (June 21, 1972), 118, no. 101: 5904–10; *NFO Reporter*, Apr. 1979.

9. Staley to Eisenhower, 1958, FF 8, DB 12, NFO Papers, ISU; Notes of Bob Casper's testimony to Senate Agriculture Committee, FF 3, DB 12, NFO Papers, ISU; NFO pamphlet, FF 25, DB 4, NFO Papers, ISU; Erhard Pfingston, NFO U.S. Farm Report, Jan. 15, 1968, film 166, NFO Papers, ISU; text of Staley speech, FF 1, DB 12, NFO Papers, ISU; *Des Moines Register*, Dec. 9, 1968; Midwest Farm Report, Oct. 8, 1964, film 7, NFO Papers, ISU; *Midlands*, Nov. 20, 1977; Richard John Gagan, "The Relation of Organizational Ideology to Role-Structure, Means, and Goals: An Analysis of the National Farmers' Organization," (masters thesis, University of Wisconsin, 1966), 54–71.

10. Galbraith, *American Capitalism*, 112–14, 154. Galbraith's notes indicate he was thinking of William H. Nicholls's *Imperfect Competition within Agricultural Industries* (Ames: Iowa State University Press, 1941). He addressed the issue of bargaining in agriculture more specifically in a Wisconsin speech in 1957. Patton, *The Case for Farmers*, 43–44; Walters, *Angry Testament*, 153; Ellis W. Hawley, *The New Deal and the Problem of Monopoly: A Study in Economic Ambivalence* (Princeton NJ: Princeton University Press, 1966), 191–97; Stephen Breyer, *Regulation and Its Reform* (Cambridge MA: Harvard University Press, 1982), 32; NFO *Reporter*, Apr. 1975.

11. Peter G. Helmberger and Sidney Hoos, "Economic Theory of Bargaining in Agriculture," *Journal of Farm Economics* 45, no. 5 (Dec. 1963): 1275–79; Robert L. Clodius, "The Role of Cooperatives in Bargaining," *Journal of Farm Economics* 39, no. 5 (Dec. 1957): 1273, 1276. The economic literature addressing bargaining is rather thin: Economists "would apparently feel more comfortable if evidence could be adduced that the gains of bargaining were at least equal to the costs, but who has the model and the method to do any such thing?" (Varden Fuller, "Bargaining in Agriculture and Industry: Comparisons and Contrasts," *Journal of Farm Economics* 45, no. 5 [Dec. 1963]: 1288); Clodius, "The Role of Cooperatives in Bargaining," 1272.

12. Wood, NFO *in Transition*, 80, 104, 110–11, 122; NFO pamphlet, FF 25, DB 4, NFO Papers, ISU; *The* NFO *Analysis of the American Livestock Marketing System*, prepared in 1961, FF 16, DB 4, NFO Papers, ISU; Brandsberg, *Two Sides in* NFO*'s Battle*, 82–83.

13. NFO pamphlet, FF 25, DB 4, NFO Papers, ISU; John T. Schlebecker, "The Great Holding Action: The NFO in September, 1962," *Agricultural History* 39 (Apr. 1965): 206–8, 210; Donata De Bruyckere, "You Can Never Say We Didn't Try: The National Farmers' Organization in Lyon County, Minnesota, 1962–1988," in *Historical Essays on Rural Life* (Marshall MN: Southwest State Univer-

sity, 1990), 16; Midwest Farm Report, Oct. 8, 1964, film 7, NFO Papers, ISU; "A Message from Your President," Dec. 1963, FF 27, DB 1, NFO Papers, ISU; American Meat Institute, Bulletin no. 1, Oct. 2, 1964, FF 9, DB 6, ser. F, Rath papers, ISU; Brandsberg, *Two Sides in NFO's Battle*, 110, 235. Unfortunately, the NFO papers do not include an accounting of the contracts signed. In 1991, when the NFO moved from Corning to Ames, seventy-five tons of materials were discarded, parts of which probably included contracts.

14. NFO pamphlet, FF 25, DB 4, NFO Papers, ISU.

15. T. W. Gloze to Oren Lee Staley, Mar. 12, 1968, and letter written on Swift stationary on Apr. 17, 1968 from Carl to Bob, FF 27, DB 7, NFO Papers, ISU; Walters, *Angry Testament*, 210; *NFO Reporter*, Feb. 1972; *NFO Reporter*, Mar. 1972; *NFO Meat Commodity-Gram*, June 8, 1973, FF 19, DB 4, NFO Papers, ISU; letters are in FFS 32, 35, 36, and 39 in DB 7 and mentioned in statements of Walt Hackney and Ed Graf to Congress, July 10, 1979, FF 17, DB 7, NFO Papers, ISU; statement of Walt Hackney, head of NFO Cattle Department, to Congress, July 10, 1979, FF 17, DB 7, NFO Papers, ISU; *Marshall Messenger*, Nov. 24, 1965; *Post-Bulletin* (Rochester MN), Mar. 24, 1966.

16. *Chicago Sun-Times*, Sept. 7, 1962; D. W. Wright to Wm. M. Cameron, Aug. 14, 1967, D. W. Wright memo, Aug. 31, 1967, and Bernard W. Ebbing to Gene Potter, May 1, 1968, all in FF 9, DB 6, ser. F, Rath papers; Preliminary Prospectus, Jan. 29, 1964, FF 12, DB 97, Financial T267, Rath Papers, ISU.

17. One article mentions that Staley outlined a block selling plan in a speech in Belgium, Wisconsin, in April 1968 (*Des Moines Register*, Apr. 14, 1968). Another article and a dissertation states that the NFO started making group sales in April 1965 (*Des Moines Register*, Apr. 13, 1968; Joseph Honan, "The National Farmers Organization: A Study of Agricultural Protest" [Ph.D. diss., University of Missouri, 1966], 169–70); Anderson, "Mission, History, and Times," 68; Harold F. Breimyer to author, Sept. 20, 1996; *The NFO Family Farmer*, FF 24, DB 4, NFO Papers, ISU; *NFO Reporter*, Feb. 1972; American Meat Institute memo, Dec. 8, 1967, FF 9, DB 6, ser. F, Rath Papers, ISU; De Bruyckere, "You Can Never Say," 15; Victor Ray to Herbert Niles, July 24, 1972, FF 41, DB 13, ser. 3, NFU Papers, UCB; "A Message from Your President," Aug. 20, 1973, FF 19, DB 4, NFO Papers, ISU; *NFO on the Move*, June 26, 1974; *NFO Grain Commodity-Gram*, June 8, 1973, FF 19, DB 4, NFO Papers, ISU; *Midlands*, Nov. 20, 1977; text of Staley speech, FF 1, DB 12, NFO Papers, ISU; NFO brochures, FF 10, DB 4, NFO Papers, ISU; *NFO on the Move*, Jan. 3, 1974.

18. Walters, *Angry Testament*, 154, 157, 159, 337, 340–41, 344; Breimyer, "Small Business Problems in the Marketing of Meat and other Commodities," Hearings before the Subcommittee on SBA and SBIC Authority and General Small

Business Problems of the Committee on Small Business, House, 96th Congress, First Session (Washington DC: GPO, 1979), 172.

19. Rowell, *Mad as Hell*, 49; *NFO Reporter*, Sept. 1975, May 1978; *NFO Reporter*, May 1978; Walters, *Angry Testament*, 158; text of Staley speech, FF 1, DB 12, NFO Papers, ISU.

20. *Des Moines Register* clipping, sometime in 1971; undated clipping, FF PL 1.1.1 National Farmers Organization 1973–74, DB 86A, MSC 414, Clark Papers, UI; Des Moines Register, Oct. 28, 1973; *NFO Reporter*, Apr. 1975; NFO to Lenders to Glenwood Packing Company, Nov. 14, 1966, FF 28, DB 11, NFO Papers, ISU; *NFO Reporter*, Aug. 1974; American Meat Institute memo, Dec. 8, 1967, FF 9, DB 6, ser. F, Rath Papers, ISU; *NFO on the Move*, May 25, 1973; *Washington Farmletter*, Jan. 8, 1971, FF 19, DB 4, NFO Papers, ISU; "Cooperation versus Coalitions" statements, FF 8, DB 7, NFO Papers, ISU; Des Moines Register, Oct. 1, 1995.

21. Herbert Hoover, "Rural Political Organizations," *South Dakota History* 13 (spring–summer 1983): 148; Torgerson, *Producer Power at the Bargaining Table*, 3–6; Ewell Paul Roy, *Collective Bargaining in Agriculture* (Danville IL: Interstate, 1970), 10–11.

22. Torgerson, *Producer Power at the Bargaining Table*, 36; S. C. Cashman, "Bargaining in Practice," *Journal of Farm Economics* 45, no. 5 (Dec. 1963): 1300.

23. Torgerson, *Producer Power at the Bargaining Table*, 36–39, 42, 44, 63, 72, 82.

24. Torgerson, *Producer Power at the Bargaining Table*, 42, 58–62, 64, 79–100, 163–64, 198–99; Walters, *Angry Testament*, 162; Donald Turner to Ernest F. Hollings, Sept. 14, 1967, and Kenneth Naden to Allen J. Ellender, Aug. 1, 1967, FF 20-246-11, Young Papers, UND; Angus McDonald, Legislative Analysis Memorandum 7-67, A Summary and More Detailed Analysis of the Background and Evolution of S. 109, Oct. 20, 1967, FF 5, DB 1, NFU–NCFC Papers, ISU.

25. Torgerson, *Producer Power at the Bargaining Table*, 54, 104–5, 119, 129–30, 132, 224; *Prairie Farmer*, June 1, 1968, 34.

26. Walter F. Mondale press release, Feb. 15, 1968, FF 20-246-24, Young Papers, UND; Tony Dechant to Walter Mondale, Feb. 16, 1968, FF 8, DB 5, NFU Papers, UCB; *Congressional Record*, 90th Cong., 2d sess. (Feb. 15, 1968), 114, pt. 3: S1288–98; *North Dakota Union Farmer* (Mar. 1968). Other cosponsors included Senators Young, Burdick, McGee, McGovern, McCarthy, Mansfield, Metcalf, Nelson, and Proxmire. The core support for the bill came from North and South Dakota, Minnesota, Wisconsin, Montana, and Wyoming. Mondale denied my requests for access to his papers housed at the Minnesota Historical Society.

27. Fuller, "Bargaining in Agriculture and Industry," 1288; Tom Huheey to Clayton Yeutter, June 5, 1972, FF McGovern, DB H, Butz Papers, PU; Baking Can-

ning Company to Gaylord Nelson, May 1, 1968, FF Agriculture: Price Bargaining, DB 258, Nelson Papers, WHS; Charles B. Shuman (AFBF President) to Allen J. Ellender, Apr. 23, 1968, FF 20-246-24, Young Papers, UND; David Angevine to Rodney Leonard, Aug. 23, 1968, FF 8, DB 5, ser. 3, NFU Papers, UCB; Robert Hampton to William Beckett, Charles Brannan, Ralph Bunje, C. L. Carpenter, Allen Mather, and John Noakes, Apr. 19, 1968, FF 8, DB 5, ser. 3, NFU Papers, UCB; *Washington Post*, Feb. 20, 1968.

28. Torgerson, *Producer Power at the Bargaining Table*, 209–10; "Suggested Areas of Possible Change in H.R. 7597," FF Congressional Correspondence, DB G, Butz Papers, PU; Richard Nixon to Keogh, Price, Buchanan, Harlow, Gavin, Anderson, Moore, Sept. 24, 1968, reprinted in William Safire, *Before the Fall* (New York: Belmont Tower, 1975), 71; *Dakota Farmer*, June 1, 1968; Jules Witcover, *85 Days: The Last Campaign of Robert Kennedy* (New York: Morrow, 1969), 191; NFO *Reporter*, Aug. 1972; Robert M. Collins, "The Economic Crisis of 1968 and the Waning of the 'American Century,'" *American Historical Review* 101 (Apr. 1996): 396–98; Clayton Yeutter to John Whitaker, Apr. 5, 1972, FF CREEP, DB G, Butz Papers, PU; *Washington Post*, Oct. 4, 1971; *Doane's Agricultural Report*, Mar. 31, 1972; NFO *Reporter*, Feb. 1974; Keith I. Clearwaters, Deputy Assistant Attorney General, "Antitrust Implications of Agricultural Bargaining," in *Bargaining Perspectives: Proceedings of the 18th National Conference of Bargaining Cooperatives* FCS Special Report 5 (Jan. 1974), 73; Thomas Kauper to Calvin Collier, Paul MacAvoy, Michael Moskow, Roger Porter, and Paul Leach, Oct. 8, 1975, FF Antitrust Immunities Task Group, DB 105, Council of Economic Advisors Papers, Ford Library.

29. NFO *Reporter*, June, Aug., and Sept. 1972; Bob Lewis to Reuben Johnson, May 10, 1978, and Lewis to John Datt, Bob Frederick, and Bob Hampton, Apr. 26, 1978, FF 20, DB 18, ser. 3, NFU Papers, UCB; Willis Rowell and Joseph Steffen, testimony prepared for congressional committees, July 1979, FF 17, DB 7, NFO Papers, ISU; M. Woodrow Wilson to Reuben Johnson, Sept. 28, 1979, FF 21, DB 18, ser. 3, NFU Papers, UCB.

30. Torgerson, *Producer Power at the Bargaining Table*, 182–83; Daniel Webster Turner to Art Thompson, Sept. 6, 1961, and Dan Turner to Mr. Wood, Jan., 5, 1961, FF NFO Correspondence 1955–61, DB 3, Turner Papers, UI. A communist was expelled from the NFO in 1976. Lee D. Sinclair to Gene Potter, Doris Peterson, Oren Lee Staley, Aug. 27, 1976, FF 24, DB 11, NFO Papers, ISU.

31. *Des Moines Register*, Mar. 16, 1956; Bourke Hickenlooper to G. W. Patterson, Aug. 28, 1967, FF NFO 1957–68, DB 36, Agriculture Committee Files, Hickenlooper Papers, Hoover Library; James Patton to Gus Geissler, Sept. 13, 1957, and Iowa NFO Board of Directors to Iowa County Officers, Dec. 24, 1957, both in FF

13, DB 7, ser. 4, NFU Papers, UCB; Warren, "The 'People's Century' in Iowa," 371–82; Pratt, "Farmers Union," 330–33; Wood, NFO in Transition, 21, 142–44; Jerome Brontmier to Gaylord Nelson, Nov. 10, 1968, FF Agriculture: Farm Organizations, DB 258, ser. M74-549, Nelson Papers, WHS; Brandsberg, *Two Sides in NFO's Battle*, 124; Lichtenstein, *Most Dangerous Man in Detroit*, 58.

32. *Des Moines Register*, Sept. 10, 1964, Mar. 19, 1967, Sept. 6 (c. 1969); unidentified clipping, FF 9, DB 9, NFO Papers, ISU; Bill Fritz to author, Aug. 14, 1996; Brandsberg, *Two Sides in NFO's Battle*, 136; Rowell, *Mad as Hell*, 29, 36; De Bruyckere, "You Can Never Say," 17–19; Wood, NFO in Transition, 142–44.

33. SAC (Special Agent in Charge), Minneapolis, to Director, FBI, Sept. 4, 1964, Oct. 16 report, and UPI wire clipping, FF 3, DB 1, NFO Papers, ISU; Roy Noonan to Karl Rolvaag and Walter Mondale, Oct. 19, 1964, and memo to Attorney General Mondale, Joseph Summers, and Paul Casey, Sept. 22, 1964, 110.F.18.1(B), Rolvaag Papers, MHS; *Des Moines Register*, Sept. 6, 1969; SAC, Milwaukee, to Director, FBI, Sept. 30, 1964, FF 3, DB 1, NFO Papers, ISU; Michigan Milk Producers Association to members, Mar. 20, 1967, FF 8, DB 7, NFO Papers, ISU; Bill Fritz to author, Aug. 14, 1996; FBI letter, Mar. 21, 1967, FF 6, DB 1, NFO Papers, ISU; Brandsberg, *Two Sides in NFO's Battle*, 129–30. Generally, see the dozens of FBI reports included in DB 1, NFO Papers.

34. *Des Moines Register*, Sept. 10, 1964; *New York Times*, Mar. 30, 1967; *Omaha World-Herald*, Mar. 31, 1967; De Bruyckere, "You Can Never Say," 9; letter of Mr. and Mrs. Bill Fritz, c. 1964, *Madison Daily Leader*, clipping in author's possession.

35. *Des Moines Register*, Mar. 19, 1968; Walters, *Angry Testament*, 322; *Waterloo Courier*, Oct. 20, 1974; A. M Beeton to Gaylord Nelson, Apr. 15, 1968, FF Agriculture: Livestock, DB 258, ser. M74-549, GNP, WHS; Jerome Brontmier to Gaylord Nelson, Nov. 10, 1968, FF Agriculture: Farm Organizations, DB 258, ser. M74-549, GNP, WHS; *New York Times*, Apr. 17, 1968.

36. Clark interview, Apr. 19, 1996; Harry Luck to Gaylord Nelson, Mar. 28, 1968, Jerome Brontmier to Gaylord Nelson, Nov. 10, 1968, FF Agriculture: Corporate Farming, DB 255, ser. M77-549, Nelson Papers, WHS.

37. Bill Fritz to author, Aug. 14, 1996; Hoover, "Rural Political Organizations," 150; W. Keith Warner and David L. Rogers, "Some Correlates of Control in Voluntary Farm Organizations," *Rural Sociology* 36, no. 3 (Sept. 1971): 327–28; De Bruyckere, "You Can Never Say," 9; Breimyer, *Policies, Attitudes and Outlook*, 6; Bourke Hickenlooper to G. W. Patterson, Aug. 28, 1967, FF NFO 1957–68, DB 36, Agriculture Committee Files, Hickenlooper Papers, Hoover Library; Victor and Evelyn Matehs to Milton Young, Feb. 24, 1968, FF 20-257-13, Young papers, UND; Gene Logsdon, *At Nature's Pace: Farming and the American Dream* (New

York: Pantheon, 1994), 8; Edward Mead and Bernhard Ostrolenk, *Voluntary Allotment: Planned Production in American Agriculture* (Philadelphia: University of Pennsylvania Press, 1933), 81–82; Brandsberg, *Two Sides of NFO's Battle*, 192, 226; Breimyer, *Over-fulfilled Expectations*, 247; Stephen Singular, "The Family Farm as a Multinational Business," *New York* 8, no. 49 (Dec. 8, 1975): 45.

38. J. Ronnie Davis and Neil A. Palomba, "The National Farmers Organization and the Prisoner's Dilemma: A Game Theory Prediction of Failure," *Social Science Quarterly* 50, no. 3 (Dec. 1969): 744, 747; George Stigler, "Free Riders and Collective Action: An Appendix to Theories of Economic Regulation," *Bell Journal of Economics and Management Science* 5 (autumn 1974): 359–60; G. W. Patterson to Tony Dechant, July 4, 1967, FF 8, DB 4, ser. 3, NFU Papers, UCB; Dick Ricci to Roger Blobaum, Feb. 9, 1968, FF 64, DB 6, ser. 3, NFU Papers, UCB; Honan, *National Farmers Organization*, 171, 175; Leighton Leon Geyer, "Farmer Bargaining: Legal, Economic, Conceptual, Theoretical, and Empirical Considerations," (Ph.D. diss., University of Minnesota, 1985), 619.

39. Rowell, *Mad as Hell*, 15; Walters, *Angry Testament*, 38; Lowell K. Dyson, *Farmers' Organizations* (New York: Greenwood, 1986), 211; copy of appeal, FF 31, DB 8, NFO Papers, ISU.

40. Dyson, *Farmers' Organizations*, 212; *Ottumwa Courier*, July 26, 1975; *SEC v. NFO*, copy of complaint in the District Court of the U.S., Southern District of Iowa, located in FF 22, DB 39, Camp Papers, UO; *State of Iowa v. NFO* injunction, FF 15, DB 8, NFO Papers, ISU; confidential NFO letter regarding SEC, June 13, 1974, FF PL 1.1.1 National Farmers Organization, 1973–74, DB 86A, MSC 414, Clark Papers, UI.

41. *New York Times*, Mar. 30, 1967, Sept. 7, 1962; Worth Rowley to Lee Sinclair, Feb. 25, 1977, FF 11, DB 2, NFO Papers, ISU; *Time*, Dec. 3, 1973; Oren Lee Staley to W. R. Poage, June 1, 1966, FF 22, DB 7, NFO Papers, ISU; *NFO Reporter*, May 1974; DeVier Pierson to LBJ, Aug. 9, 10, 1967, LBJ Library.

42. Walters, *Holding Action*, 22, 24–25; Iowa NFO BOD to Iowa County Officers, Dec. 24, 1957, FF 13, DB 7, ser. 4, NFU Papers, UCB; Brandsberg, *Two Sides in NFO's Battle*, 209–16, 219; Rowell, *Mad as Hell*, 88–95; Charles Walters, "No Wish to Dispute the Fuehrer," *Key to the News* (Kansas City–St. Joseph), Dec. 24, 1972; Clara B. Riveland, "An Analysis of the National Farmers Organization's Attempts to Reduce Rhetorical Distance" (Ph.D. diss., University of Minnesota, 1974), 306–13.

43. Brandsberg, *Two Sides of NFO's Battles*, 233; *Des Moines Register*, Sept. 1, 1982; Farm Bureau, "Discussion of Cooperative Marketing Problems and Developments," Aug. 23, 1963, FF 8, DB 7, NFO Papers, ISU; James Patton to Monsignor O'Rourke, Oct. 6, 1964, FF 16, DB 7, ser. 4, NFU Papers, UCB; Hubert Humphrey

to James Patton, Apr. 5, 1961, and Humphrey to Reverend James Vizzard, Apr. 5, 1961, FF 16, DB 7, ser. 4, NFU Papers, UCB; Mrs. E. Rohl to Milo Swenton, Sept. 3, 1963, FF NFO, DB 8, and Owen Hullberg to Orville Freeman, Oct. 27, 1962, FF NFO 1962–63, DB 14, both in M71-218, Wisconsin Federation of Cooperatives Papers, WHS; William F. Thompson, *The History of Wisconsin*, vol. 4, *Continuity and Change, 1940–1965* (Madison: State Historical Society of Wisconsin, 1988), 156; Wood, *NFO in Transition*, 138; Thomas Kauper (Assistant Attorney General, Antitrust Division) to Calvin Collier, Paul MacAvoy, Michael Moscow, Roger Porter, Paul Leach, Oct. 8, 1975, FF Antitrust Immunities Task Group, DB 105, Council on Economic Advisors Papers, Ford Library.

6. FARMER COOPERATIVE MARKETING

1. James Patton to the NFU Board of Directors, Sept. 8, 1961, FF 7-25 NFU Board Meetings, 1960–66, DB 7, ser. 4, NFU Papers, UCB; *Farmers' Union Save Our Coop* report, FF 9-18, Save Our Coop Campaign, DB 9, ser. 2, NFU Papers, UCB; NCFC quoted in Laszlo Valko, *Cooperative Laws in the USA: Federal Legislation, 1890–1980*, Bulletin 0902, 1981, Washington State University, College of Agriculture, Research Center, 19; Harold Breimyer, "Farms, Farmers, and Farm Policy in an Industrial Age," speech to Minnesota Seminar in Public Affairs, Jan. 16, 1962, mailed to author; William Heffernan and Douglas Constance, "Transnational Corporations and the Globalization of the Food System," in Allesandro Bonanno, Lawrence Busch, William Friedland, Lourdes Gouveia, and Enzo Mingione, eds., *From Columbus to ConAgra: The Globalization of Agriculture and Food* (Lawrence: University Press of Kansas, 1994), 34; Roy, *Contract Farming and Economic Integration*, 166–68.

2. Austin T. Flett to Paul Douglas, Apr. 7, 1953, FF 17, DB 8, M63-014, Cooperative League Papers, WHS; William B. Baum to Arthur Capper, May 21, 1949, FF Agri. Correspondence, DB 34, Capper Papers, KSHS.

3. "Cooperative Growth: Trends, Comparisons, Strategy," FCS Information 87, Farmer Cooperative Service, USDA, Mar. 1973, 52–54; Mack Leath, Lowell Hill, Bruce Marion in Marion, ed., *Organization and Performance of the U.S. Food System*, 156.

4. Brinkley, *End of Reform*, 3–8; Louis Galambos and Joseph Pratt, *The Rise of the Corporate Commonwealth: U.S. Business and Public Policy in the Twentieth Century* (New York: Basic Books, 1988), 129–37, 153–54; Alan Wolfe, *America's Impasse: The Rise and Fall of the Politics of Growth* (New York: Pantheon, 1981), 10–11, 15–26; Robert Griffith, "Forging America's Postwar Order: Domestic Politics and Political Economy in the Age of Truman," in Michael J. Lacey, *The Truman Presidency* (New York: Cambridge University Press, 1989), 74–75, and

"Dwight D. Eisenhower and the Corporate Commonwealth," *American Historical Review* 87 (Feb. 1982): 107; Robert M. Collins, *The Business Response to Keynes, 1929–1964* (New York: Columbia University Press, 1981), 129–52; Kim McQuaid, *Big Business and Presidential Power: From FDR to Reagan* (New York: Morrow, 1982), 122–49; Gordon, *New Deals,* 294–305; Lichtenstein, "From Corporatism to Collective Bargaining," 122–45; Maier, *In Search of Stability,* 125–30; Elizabeth Fones-Wolf, *Selling Free Enterprise: The Business Assault on Labor and Liberalism, 1945–1960* (Urbana: University of Illinois Press, 1994), 2.

5. Richard Sexton, "Increased Competition in Agricultural Markets and the Role of Cooperatives," *American Journal of Agricultural Economics* 72 (Aug. 1990), 709–19.

6. Lawrence Goodwyn, *The Populist Moment: A Short History of the Agrarian Revolt in America* (New York: Oxford University Press, 1978), vii, 30, 34, 66, 74, 83, 87, 89, 130; Gilbert C. Fite, *Farm to Factory: A History of the Consumers Cooperative Association* (Columbia: University of Missouri Press, 1965), 5–9; Steven L. Piott, *The Anti-Monopoly Persuasion: Popular Resistance to the Rise of Big Business in the Midwest* (Westport CT: Greenwood, 1985), 16–21.

7. Theodore Saloutos and John D. Hicks, *Agricultural Discontent in the Middle West, 1900–1939* (Madison: University of Wisconsin Press, 1951), 286; Grace H. Larsen and Henry E. Erdman, "Aaron Sapiro: Genius of Farm Co-operative Promotion," *Mississippi Valley Historical Review* (Sept. 1962): 242–68 (Sapiro quoted on p. 253); Joan Hoff Wilson, "Hoover's Agricultural Policies, 1921–1928," *Agricultural History* 51 (Apr. 1977): 359.

8. James L. Guth, "Farmer Monopolies, Cooperatives, and the Intent of Congress: Origins of the Capper-Volstead Act," *Agricultural History* 56 (Jan. 1982): 67–82 (quote from p. 81); Wilson, "Hoover's Agricultural Policies," 335–61 (quote from p. 337); Hamilton, *From New Day to New Deal,* 33.

9. Larsen and Erdman, "Aaron Sapiro," 251; Elizabeth Hoffman and Gary Libecap, "Institutional Choice and the Development of U.S. Agricultural Policies in the 1920s," *Journal of Economic History* 51 (June 1991): 397–411; Fite, *George N. Peek,* 135; William Ellis, "Robert Worth Bingham and the Crisis of Cooperative Marketing in the Twenties," *Agricultural History* 56 (Jan. 1982): 99–116; Glenna Matthews, "The Apricot War: A Study of the Changing Fruit Industry during the 1930s," *Agricultural History* 59 (Jan. 1985): 37; Fite, *American Farmers,* 33; Hamilton, *From New Day to New Deal,* 13–14.

10. Hamilton, *From New Day to New Deal,* 35–39, 43–44.

11. Hamilton, *From New Day to New Deal,* 47, 50–51, 57.

12. Hamilton, *From New Day to New Deal,* 59–60, 63–64, 75, 79–80, 82, 84–85, 106. The Farmers' Union was divided between supporters of cooperatives and

supporters of government intervention. Tony Dechant speech to 51st National Institute of Cooperation, Aug. 7, 1979, FF 19, DB 19, ser. 3, NFU Papers, UCB.

13. Hamilton, *From New Day to New Deal*, 132–34, 136–38, 142; *Aberdeen American News*, Mar. 8, 1959; Bush, "Political Economy of Wheat," 252.

14. Hamilton, *From New Day to New Deal*, 232; Joseph Knapp, *The Advance of American Cooperative Enterprise, 1920–1945* (Danville IL: Interstate, 1973), 232, 245, 252–53, 255, 263, 265, 501; "Cooperative Development Project," 1966, FF 11, DB 2, ser. 3, NFU Papers, UCB. The functions of the older Division of Cooperative Marketing, established in the 1920s, were also continued.

15. Goodwyn, *Populist Moment*, 111; Knapp, *Advance of American Cooperative*, 403–7, 494–95; Jerry Voorhis, *American Cooperatives* (New York: Harper & Brothers, 1961), 86; Jane Scearce, "What's Doing in Our State Councils," Farm Credit Administration paper, FF FCA, 1948, DB 2, 143.E.11.2(F), Minnesota Association of Cooperatives Papers, MHS; Harry Peterson to M. W. Thatcher, Jan. 24, 1950, FF FUGTA, 1950, DB 9, 143.E.11.9(B), Minnesota Association of Cooperatives Papers, MHS; "Report of National Council Organization Committee," Apr. 7, 1964, FF 7, DB 21, Iowa Institute of Cooperation Papers, ISU.

16. Knapp, *Advance of American Cooperative*, 415–16, 425–26, 486, 495; pamphlet, "The Farmers Answer to 'Monopoly in Action,'" Farmers Union Grain Terminal Association, 1947, available from MHS; Arthur Capper to George C. Marshall, Aug. 30, 1948, FF Agricultural Correspondence, DB 34, Capper Papers, KSHS.

17. Arkansas Wholesale Grocers' Association, Bulletin no. 149, Oct. 10, 1944, FF Agric. Correspondence, Cooperatives: Flour, DB 35, Capper Papers, KSHS; Knapp, *Advance of American Cooperative*, 521–24, 537; Milton D. Hakel, *Who Needs a Farmers Union—And Why?*, 20, DB 7, ser. 2, NFU Papers, UCB; Glenn Edick, Royce Jordan, Albert Ortego, Gerald Pepper, Ronald Schuler, Dempsey Seastrunk, James Schaffer, Melvin Sims, Gene Swackhammer, George Thomas, "A Review and Evaluation of the ESCS Cooperatives Program," Feb. 15, 1980, 4, FF Cooperatives, DB 119, M80-626, Nelson Papers, WHS; unpublished manuscript about the GTA, 214–15, FF History of FUGTA in 2, DB 1, 149.E.10.6(F), Thatcher Papers, MHS (hereinafter GTA manuscript); NTEA, Bulletin 36, Nov. 19, 1945 and Bulletin 24, Feb. 9, 1945, FF Agric. Correspondence, Cooperatives: Flour, DB 35, Capper Papers, KSHS; Davis Doubitt, "Taxes and Co-ops," Midland Cooperative Wholesale, FF Cooperatives, 1947–51, DB 1, 144.K.7.3, Freeman Papers, MHS.

18. NTEA, "Cooperative Expansion in the Petroleum Industry," 1944, i, 2, 4–5, and Ron Kennedy to All Peavey System Employees in the U.S., May 22, 1958, both in FF Cooperative Associations: Opposition Counterattacks, undated 1944–81, DB 4, 145.L.9.4(F), CENEX Papers, MHS; Rogers K. Rose to Len Jordan, Mar. 2, 1967, FF 20-246-11, Young Papers, UND.

19. R. B. Laing to Arthur Capper, Oct. 21, 1944, and Arkansas Wholesale Grocers' Association, Bulletin no. 149, Oct. 10, 1944, FF Agric. Correspondence, Cooperatives: Flour, DB 35, Capper Papers, KSHS; Roswell Magill, "Taxation of the Income of Cooperatives," speech to Grain and Feed Dealers Association, Jan. 23, 1958, DB 29, 145.E.12.2(F), McCarthy Papers, MHS; NTEA, "How Cooperatives Escape the Income Tax," FF Cooperatives, 1947–51, DB 1, 144.K.7.3, Freeman Papers, MHS.

20. GTA manuscript, 212, 216; "The Taxation of Farmer Cooperatives," NCFC, July 11, 1945, and Henry Jackson press release, June 18, 1948, both in FF Agric. Correspondence, Cooperatives: Flour, DB 35, Capper Papers, KSHS; Arthur Capper, "Farmer Cooperatives' Views on Taxation," *Congressional Record*, 93d Cong., 1st sess. (Feb. 19, 1947), 80, pt. 1: 1154; Davis Doubtit, "Taxes and Co-ops," Midland Cooperative Wholesale, FF Cooperatives, 1947–51, DB 1, 144.K.7.3, Freeman Papers, MHS.

21. Poem cited in GTA manuscript, 207.

22. John Earl Haynes, "Farm Coops and the Election of Hubert Humphrey to the Senate," *Agricultural History* 57 (Apr. 1983): 202–5; Harry Peterson to M. W. Thatcher, Oct. 17, 1947, FF FUGTA, 1947, DB 2, 143.E.11.2(F), Minnesota Association of Cooperatives Papers, MHS; Thomas Stokes, "War in GOP on Co-ops," editorial in unidentified paper, FF Agric. Correspondence, Cooperatives: Flour, DB 35, Capper Papers, KSHS; Harold Knutson to the Voters of the Sixth District, June 6, 1946, FF Politics, 1946, DB 1, 143.E.11.1(B), Minnesota Association of Cooperatives Papers, MHS; *Midland Cooperator*, Nov. 27, 1946; Bob Hendschin to M. W. Thatcher, Aug. 14, 1948, FF FUGTA, DB 5, 143.E.11.5(B), Minnesota Association of Cooperatives Papers, MHS; NTEA Bulletin 97, Feb. 21, 1949, FF NTEA, 1949, DB 7, 143.E.11.7(B), Minnesota Association of Cooperatives Papers, MHS. In 1946 the House Small Business Committee issued a report favoring cooperatives and reaffirmed it in 1947. *Washington Situation* (NCFC) 8, no. 47 (Nov. 21, 1947), FF Agric. Correspondence, Cooperatives: Flour, DB 35, Capper Papers, KSHS; Wisconsin Association of Cooperatives, July 9, 1957, FF 8, DB 5, pre-1957 collection, Proxmire Papers, WHS.

23. Newspaper clipping, "Humphrey Hails His Friend Bill," FF Biographical Information, 149.E.10.6(F), Thatcher Papers, MHS; GTA *Gleaner* 7, no. 8 (Dec. 1968), MHS; GTA manuscript, 224–26; Ron Kennedy to All Peavey System Employees in the U.S., May 22, 1958, FF Cooperative Associations: counterattacks, undated and 1944–81, DB 4, 145.L.9.4(F), CENEX Papers, MHS; Jerry Voorhis to M. W. Thatcher, May 19, 1947, FF 11, DB 20, H63-014, Cooperative League of USA Papers, WHS; *Business Week*, Feb. 7, 1977, 64; Haynes, "Farm Coops and the Election of Hubert Humphrey to the Senate," 209–10; Thomas G. Ryan, "Farm Prices

and the Farm Vote in 1948," *Agricultural History* 54 (July 1980): 389; Virgil W. Dean, "The Farm Policy Debate of 1949–50: Plains State Reaction to the Brannan Plan," *Great Plains Quarterly* 13 (winter 1993): 36.

24. James Patton to Dwight Eisenhower, Sept. 9, 1952, FF 15, DB 10, James G. Patton Papers, UCB; Eisenhower quoted by Senator Frank Carlson (R, KS) in a speech to the Grain Terminal Association, Dec. 1952, FF Corporate Records and Annual Meetings, 1952, DB 2 149.E.10.7 (B), Thatcher Papers, MHS.

25. Memorandum on H.J. Res. 591 and 592, S.J. Res. 172, FF 142 H Cooperatives, DB 1151, GF 142-6-19, Eisenhower Library; Thatcher to Barron County (WI) Farmers Union, Sept. 1973, FF Speeches by Thatcher, 1955–73, DB 4, 149.E.10.9(B), Thatcher Papers, MHS; Farmers Union Jobbing Association Annual Report, 1959, FF 7, DB 2, Far-Mar-Co Papers, KSU; *Aberdeen American News*, Mar. 8, 1959; newspaper clipping, "Co-op Tax Fight Was One of Thatcher's Brilliant Victories," FF Biographical Information, 149.E.10.6(F), Thatcher Papers, MHS.

26. C. O. Ryde to Arthur Capper, June 24, 1948, FF Agric. Correspondence, Cooperatives: Flour, DB 35, Capper Papers, KSHS; Joseph G. Knapp, "Farmer Cooperative Service Gets Under Way," *News for Farmer Cooperatives* (FCS, May 1954); "Farmer Cooperative Service Purpose Statement," 3, FF 27, DB 24, M69-268, Wisconsin Federation of Cooperatives Papers, WHS; D. H. McVey and William Summitt, "Cooperatives Extend Reach of Grain Producers," *News for Farmer Cooperatives* 31, no. 10 (FCS, Jan. 1965): 16; Fred Merrifield, "How Do Banks for Cooperatives Loans Fit into Growth and Expansion Programs for Farmer Cooperatives?" in NCFC press release, Jan. 13, 1959, FF 6, DB 21, Iowa Institute of Cooperation Papers, ISU.

27. P. J. Nash to Frank Carlson, Aug. 11, 1959, FF Legislative Correspondence 1959–60, DB 172, Carlson Papers, KSHS; James G. Patton to Dwayne O. Andreas, Sept. 7, 1961, FF 23, DB 2, ser. 4, NFU Papers, UCB; Angus McDonald to Ray Obrecht, May 9, 1968, FF 5-8 Agriculture Farm Bargaining and Marketing, DB 5, ser. 3, NFU Papers, UCB; Alfred Stedman, "Above the Law?" *St. Paul Pioneer Press*, Aug. 30, 1959; Farmers for Kennedy-Johnson press release, FF 2, DB 5, Iowa Institute of Cooperation Papers, ISU; James Patton to the NFU Board of Directors, Sept. 8, 1961, FF 7-25 NFU Board Meetings, 1960–66, DB 7, ser. 4, NFU Papers, UCB; White House press release, Jan. 31, 1964, FF 17, DB 10, Patton Papers, UCB.

28. *News for Farmer Cooperatives* 31, no. 10 (Jan. 1965): 9; "Cooperative Development Project," 1966, FF 11, DB 2, ser. 3, NFU Papers, UCB; Office of Economic Opportunity press release, FF Pig Marketing Cooperative, DB 303, M74-549, Nelson Papers, WHS; "The Application of Antitrust Laws to Certain Mergers of Cooperatives," FF 23, DB 2, ser. 4, NFU Papers, UCB; Division of Community De-

velopment Services, "New Opportunities for Farmer Cooperatives," Dec. 15, 1964, FF 21, DB 1, ser. 3, NFU Papers, UCB; James G. Patton to R. Sargent Shriver, Apr. 3, 1967, FF 34, DB 4, ser. 3, NFU Papers, UCB. The story of cooperative promotion during the War on Poverty is left out of Allen Matusow's *The Unraveling of America: A History of Liberalism in the 1960s* (New York: Harper Torchbooks, 1984).

29. Roy Hendrickson to Co-op Friends, July 21, 1961, FF 5, DB 22, Iowa Institute of Cooperation Papers, ISU; Robert Anderson to Carl T. Curtis, FF 17, DB 56, Far-Mar-Co Papers, KSU; Donald R. Timmel to Abe and Sam Cohen, July 1, 1977, FF 16, DB 6, Far-Mar-Co Papers, KSU; *Wichita Eagle and Beacon,* July 30, 1977.

30. Nebraska Farmers Union to Bourke Hickenlooper, Mar. 8, 1956, FF NFU 1950–56, DB 36, Agriculture Committee Files, Hickenlooper Papers, Hoover Library; Robert P. Combs and Bruce W. Marion, "Food Manufacturing Activities of 100 Large Agricultural Marketing Cooperatives," WP-73, Apr. 1984, 1, 6, 13, 18–19, 21, 24, 28, 30, 32, 36, 40, 46, 51.

31. "Market Structure and Pricing Efficiency of U.S. Grain Export System," General Accounting Office CED-82-61, June 15, 1982, 17; NCFC press releases, Mar. 2, Nov. 12, 1970, FF 10, DB 1, NFU–NCFC Papers, ISU; Kenneth Naden to Maurice Stans, June 9, 1969, and NCFC, Farmers' Union, Farm Bureau, Grange to Nixon, telegram, Mar. 5, 1969, both in FF 8-7, NCFC, DB 8, ser. 3, NFU Papers, UCB; Everett Dirksen to Tony Dechant, Apr. 5, 1968, FF Foreign Trade Exports and Imports, DB 5, ser. 3, NFU Papers, UCB; NCFC, Grange, NFU, Farm Bureau, NFO, National Federation of Grain Cooperatives, National Association of Corn Growers, National Association of Wheat Growers, Western Wheat Associates, Midcontinent Farmers Association, American Soybean Association, Grain Sorghum Producers Association, and Great Plains Wheat to Nixon, Nov. 12, 1970, FF 17, DB 10, ser. 3, NFU Papers, UCB.

32. Farmers Union Marketing and Processing Association, 47th Annual Report, 1976, FF Corporate Records 3, DB 4, 1974–76, 149.E.10.9 (B), Thatcher Papers, MHS; Division of Community Development Services, "New Opportunities for Farmer Cooperatives," Dec. 15, 1964, FF 21, DB 1, ser. 3, NFU Papers, UCB; James Patton, Action Report: Farmers Union Commission on Cooperation, May 1961, FF Cooperatives: Farmers Union Commission on, 1961, DB 1, ser. 3, NFU Papers, UCB; Willard Cochrane to Orville Freeman, Mar. 20, 1964, FF 1964 W. W. Cochrane memoranda (1), DB 11, 144.K.8.5(B), Freeman Papers, MHS; Tony Dechant to Bob Lewis, Nov. 12, 1973, and FUMPA press release, July 2, 1973, FF 7, DB 16, ser. 3, NFU Papers, UCB.

33. Minutes, joint meeting Boards of Directors and Management of GTA and CENEX, and NFU President, Apr. 4, 1974, FF 26, DB 22, ser. 3, NFU Papers, UCB;

Terry Murphy to Managers and Directors of Cooperatives in the State of Montana, Nov. 4, 1976, and press release, July 30, 1976, FF 14, DB 30, ser. 3, NFU Papers, UCB; *Montana Farmers' Union Newsletter*, Jan. 1980.

34. Gilbert C. Fite, *Beyond the Fence Rows: A History of Farmland Industries, Inc., 1929–1978* (Columbia: University of Missouri Press, 1978), vii, 135–36, 160, 181–82, 184–91, 220, 222, 230, 249, 280–82.

35. Fite, *Beyond the Fence Rows*, 260–61, 276, 306–7, 313.

36. Fite, *Beyond the Fence Rows*, 308–12, 314; "Cooperative Growth: Trends, Comparisons, Strategy," FCS Information 87, Farmer Cooperative Service, USDA, Mar. 1973, 39.

37. Thatcher to Barron County (WI) Farmers Union, Sept. 1973, FF Speeches by Thatcher, 1955–73, DB 4, 149.E.10.9(B), Thatcher Papers, MHS; Mr. Strong and Mr. Frazee to Mr. Sanders, June 10, 1943, FF Annual Meeting, 1943, DB 1, Thatcher Papers, MHS; Charles Brannan to M. W. Thatcher, Mar. 29, 1960, FF 54, DB 3, ser. 4, NFU Papers, UCB; FUGTA, *Ninth Annual Report, 1946*, 11, FF Corporate Research, Annual Meeting 1946, DB 1, Thatcher Papers, MHS; GTA manuscript, 216; *Aberdeen American News*, Mar. 8, 1959.

38. Biographical sketch, M. W. Thatcher, Mar. 1, 1967, FF 3-31 Farmers Union Grain Terminal Association, 1967, DB 3, ser. 3, NFU Papers, UCB; Thatcher manuscript, "Farm Prices Are Made in Washington," FF Biographical Information 1931–64, DB 1, Thatcher Papers, MHS.

39. Mr. Strong and Frazee to Mr. Sanders, June 10, 1943, and newspaper clipping, "Story of Thatcher Career Reads Like a Novel," FF Annual Meeting, DB 1, FF Biographical Information 1931–64, Thatcher Papers, MHS; Thatcher to Tony Dechant, June 11, 1968, FF FUGTA, DB 5, ser. 3, NFU Papers, UCB; O. K. Armstrong, "Why Should These Co-ops Enjoy Special Tax Privileges?" *Reader's Digest*, Feb. 1962, 71.

40. "Story of Thatcher Career Reads Like a Novel"; *Minneapolis Tribune*, Feb. 20, 1966; M. W. Thatcher to Richard Johansen, Mar. 1, 1966, FF Andreas Brothers Contract re: ADM 1 1966, 149.E.10.(F), Thatcher Papers, MHS; *Minneapolis Tribune*, Jan. 4, Feb. 20, 1966; *Farmers Union Herald*, Jan. 10, 1966.

41. Bill Wagner, "The Widening Horizons of the GTA," *Corporate Report*, Sept. 1976, 35; GTA news release, n.d., FF 15, DB 24, ser. 3, NFU Papers, UCB.

42. George Bicknell to Robert Lunsford, May 8, 1968, FF 13, DB 2, Far-Mar-Co Papers, KSU; *Leadership*, July 1968, 11; "Persons in Attendance at Special Meeting at the Wichita Bank for Cooperatives for Discussing Mergers," Oct. 28, 1965, FF 9, DB 45, Far-Mar-Co Papers, KSU; Roy Hendrickson, "MWT Combined Business Genius with Love for Those on the Land," *Farmers Union Herald* 42, no. 11 (June 3, 1968).

43. "Far-Mar-Co–Farmland Merger Position Paper," 2, FF 2, DB 46, Far-Mar-Co Papers, KSU; George Voth to Earnest Lindsey, May 22, 1978, FF 3, DB 39, Far-Mar-Co Papers, KSU; "Meeting to Discuss Possible Purchase of ADM Properties," May 19, 1972, FF 6, DB 33, Far-Mar-Co Papers, KSU; H. C. Clark to E. T. Lindsey, June 13, 1972, FF 6, DB 33, Far-Mar-Co Papers, KSU; Fite, *Beyond the Fence Rows*, 363.

44. "Far-Mar-Co Grain Marketing Procedures," FF 19, DB 53, Far-Mar-Co Papers, KSU; "Far-Mar-Co Subterminal Feasibility Studies," Oct. 13, 1980, FF 4, DB 53, Far-Mar-Co Papers, KSU; "Superior, Nebraska, Grain Subterminal Study," Sept. 1980, FF 9, DB 53, Far-Mar-Co Papers, KSU.

45. Ronald Knutson, "Cooperatives Competitive Position in the Grain Industry," speech to the Farmers Grain Dealers Association of Iowa, Jan. 28, 1975, FF 22, DB 56, Far-Mar-Co Papers, KSU; Fite, *Beyond the Fence Rows*, 370; Bill Wagner, "The Widening Horizons of the GTA," *Corporate Report*, Sept. 1976, 38; "Far-Mar-Co–Farmland Merger Position Paper," 2, FF 2, DB 46, Far-Mar-Co Papers, KSU; *News for Farmer Cooperatives*, Mar. 1976, 15; Report of the Promark Implementation Committee, May 16, 1975, FF 13, DB 56, Far-Mar-Co Papers, KSU. The 1977 level was 37 million bushels. *Promark News* 1, no. 2 (Sept. 1978), DB 68, Far-Mar-Co Papers, KSU.

46. "Promark 1978 Final Settlement Report," FF 6, DB 60, Far-Mar-Co Papers, KSU; "Survey of Promark Participants," Oct. 1979, Farmland Economic and Market Research, FF 2, DB 53, Far-Mar-Co Papers, KSU.

47. *Minneapolis Tribune*, Jan. 21, 1979; "Cooperative Growth: Trends, Comparisons, Strategy," FCS Information 87, Farmer Cooperative Service, USDA, Mar. 1973, 51; Bill Wagner, "The Widening Horizons of the GTA," *Corporate Report*, Sept. 1976, 38; "Story of Thatcher Career Reads Like a Novel"; pamphlet, *GTA: A Chronology, 1938–1976*, available at MHS; Dwayne Andreas to M. W. Thatcher, Feb. 24, 1966, FF Andreas Brothers Contract re: ADM 1 1966, 149.E.10.(F), Thatcher Papers, MHS; "Far-Mar-Co–Farmland Merger Position Paper," 2; Goldman, Sachs, and Co., "Rating Agency Results and Financing Proposal for the Port of Galveston Grain Elevator: Report to the Board of Directors Farmers Export Company," Aug. 20, 1976, FF 13, DB 60, Far-Mar-Co Papers, KSU.

48. James G. Patton to Dwayne O. Andreas, Sept. 7, 1961, FF 23, DB 2, ser. 4, NFU Papers, UCB; Fite, *Beyond the Fence Rows*, 289, 300, 363, 369; "Far-Mar-Co–Farmland Merger Position Paper," 2; *Business Week*, Aug. 4, 1980, and July 23, 1984; press release, "New Farm Co-op Unveiled at News Conference, June 1," FF GTA 1966, 1978–82, DB 6, 147.K.11.7(B), MAC Papers, MHS; *Kansas City Times*, Oct. 15, 1996; Bush, "Political Economy of Wheat," 279–81.

49. *Farmers Corn Products, Inc. (FCPI) Newsletter*, no. 9 (Sept. 16, 1981); FCPI,

Questions and Answers; FCPI Fact Sheet, all in FF Farmers Corn Products, Inc. 1980–82, DB 6, 147.K.11.7(B), MAC Papers, MHS; Dan Looker, "Strength in Numbers," *Successful Farming*, Nov. 1994, 20–22; *Los Angeles Times*, Apr. 9, 1995; phone interview with Doug Leet, director of Farm Service Agency, Waseca County, Jan. 29, 1997; Minnesota Corn Processors pamphlet, in author's possession.

50. Bob Beasley to Earnest Lindsey, John Anderson, Don Ewing, Bob Johanson, Gordon Leith, Ken Neilsen, George Voth, and Dave Fulton, Oct. 6, 1977, FF 4, DB 39, Far-Mar-Co Papers, KSU; Gaylord Nelson to Bob Bergland, Apr. 25, 1980, Rod Nilsestuen to Jeff Nedelman, Apr. 22, 1980, and Kenneth Naden to Howard W. Hjort, Apr. 3, 1980, all in FF Cooperatives, DB 119, M80-626, Nelson Papers, WHS; Dear Colleague from Robert Morgan, Mar. 22, 1979, FF S. Res 136, DB 119, M80-626, Nelson Papers, WHS.

51. Edick et al., "Review and Evaluation."

52. Tony Dechant to State Farmers Union Presidents, Aug. 20, 1969, FF 39, DB 9, ser. 3, NFU Papers, UCB; Fite, *Beyond the Fencerows*, 348–49; Garner M. Lester to Loren Horst, Jan. 12, 1981, FF Cooperative Associations: counterattacks, undated and 1944–81, DB 4, 145.L.9.4(F), CENEX Papers, MHS; *Business Week*, Feb. 7, 1977, 57, 64; B. J. Malusky to All GTA Elevator Presidents, Sept. 18, 1969, FF 61, DB 7, ser. 3, NFU Papers, UCB. Questions about the antitrust status of cooperatives are summarized by Bruce Bohlman, "Agricultural Cooperatives and the Search for Parity: A Confrontation with the Antitrust Laws," *North Dakota Law Review* 44, no. 4 (summer 1968): 425–552.

53. "Washington Report from NTEA," Oct. 22, 1980, FF Cooperative Associations: counterattacks, undated and 1944–81, DB 4, 145.L.9.4(F), CENEX Papers, MHS; Hubert Humphrey to Milt Hakel, July 14, 1973, FF Farmers Union, DB 150.J.6.9(B), Humphrey Papers, MHS; Kenneth Naden to NCFC Members and Board, July 26, 1974, FF Earl Butz: Antitrust Speech, 1976, DB 4, 152.C.7.3(B), MAC Papers, MHS; Shover, *First Majority–Last Minority*, 188; Dear Colleague from Ed Mezvinsky, Sept. 12, 1975, FF Judiciary: Food Industry, DB 47, Culver Papers, UI; Edick et al., "Review and Evaluation," 5; *Business Week*, Feb. 7, 1977, 55; Fite, *Beyond the Fencerows*, 370–74.

54. "Minutes of Meeting of Task Force on Antitrust and Monopoly Problems," Aug. 14, 1974, and press release, John Heinz, Aug. 23, 1974, FF 12, DB 30, 1974, Agribusiness Accountability Project Papers, ISU; Farmers Union press release, Mar. 25, 1975, FF 27, DB 22, ser. 3, NFU Papers, UCB; *Wall Street Journal*, Oct. 15, 1973; Associated Press, Nov. 6, 1975, FF Earl Butz: Antitrust Speech, 1976, DB 4, 152.C.7.3(B), Minnesota Association of Cooperatives Papers, MHS; "The Sorry Republican Record in Agriculture," 13, DB 83, 149.H16.9(B), Bob Bergland Papers, MHS; "A Report on Agricultural Cooperatives," FF Food Deputies Group,

1975–76, DB 108, Council of Economic Advisers Papers, Ford Library; Robert Lewis to Peter Rodino, Mar. 16, 1976, FF 27, DB 22, ser. 3, NFU Papers, UCB; James E. Anderson, "Agricultural Marketing Orders and the Process and Politics of Self-Regulation," *Policy Studies Review* 2, no. 1 (Aug. 1982): 106–7.

55. Truman Graf, "Agricultural Bargaining: Forward or Backward?" *Economic Issues*, no. 35 (July 1979), Department of Agricultural Economics, College of Agricultural and Life Sciences, University of Wisconsin-Madison; Tony Dechant speech to 51st National Institute of Cooperation, Aug. 7, 1979, FF 19, DB 19, ser. 3, NFU Papers, UCB.

56. Graf, "Agricultural Bargaining"; "Statement of NFU Concerning USDA Review of Rules of Practice Governing Cease and Desist Proceedings under Section 2 of the Capper Volstead Act," July 6, 1979, FF 28, DB 22, ser. 3, NFU Papers, UCB; Bob Lewis to Reuben Johnson, May 10, 1978, FF 20, DB 18, ser. 3, NFU Papers, UCB; Bob Bergland to the National Commission for the Review of Antitrust Laws and Procedures, July 27, 1978, FF 18, DB 9, ser. 2, NFU Papers, UCB; *Coop Country News*, Mar. 1979; see flyers, FF 18, DB 9, ser. 2, NFU Papers, UCB.

57. Jerry Voorhis to Wright Patman, Feb. 15, 1955, FF 40, DB 16, M63-014, Cooperative of USA Papers, WHS; Edward Slettom to Donald Fraser, Mar. 2, 1976, FF Oil Divestiture 1976–77, 1, and Slettom to Hubert Humphrey, July 7, 1976, FF Oil Divestiture 1976–77, 2, both in DB 5, 152.C.7.4(F), Minnesota Association of Cooperatives Papers, MHS; Thurman Arnold to Jerry Voorhis, July 25, 1949, FF 44, DB 3, M63-014, Cooperative League Papers, WHS; Reuben Johnson to House members, Oct. 26, 1979, FF 28, DB 22, ser. 3, NFU Papers, UCB.

58. Victor Ray to Tony Dechant and Bob Lewis, Dec. 15, 1977, FF 27, DB 22, ser. 3, NFU Papers, UCB; "An Exploratory Meeting Attended by American Agriculture Movement Leadership and Farmland–Far-Mar-Co Leadership," Nov. 27, 1978, FF 13, DB 32, Far-Mar-Co Papers, KSU; "Developing New Subterminals: Case Studies and Guidelines," FF 3, DB 53, Far-Mar-Co Papers, KSU; Edick et al., "Review and Evaluation," 11; John Love to Mr. Hallberg, Mar. 7, 1977, FF 2, DB 46, Far-Mar-Co Papers, KSU; Fite, *Beyond the Fence Rows*, 364–67; Linda Kravitz, *Who's Minding the Coop?* (Washington DC: Agribusiness Accountability Project, 1974), 17; Randall E. Torgerson, "Farm Bargaining: Some Practical Considerations," 22, Agricultural Economics Paper 1971-26, University of Missouri "Think Tank on Farm Bargaining," Mar. 1971, FF 62, DB 11, ser. 3, NFU Papers, UCB; Victor Ray to Tony Dechant and Robert Lewis, Feb. 11, 1974, and Bob Lewis to Tony Dechant, Jan. 2, 1974, FF 20, DB 18, ser. 3, NFU Papers, UCB.

59. W. E. Black, "Cooperative Grain Marketing Today and Tomorrow," speech to Sixth Annual Meeting of Far-Mar-Co, Feb. 5, 1974, FF 27, DB 22, ser. 3, NFU Papers, UCB; "USDA Team Tells Cooperatives How to Improve Grain Marketing,"

FF 14, DB 56, Far-Mar-Co Papers, KSU; Knapp, "Cooperative Expansion through Horizontal Integration," FF 1950s, DB 1, and "Competition between Cooperatives: Meeting the Problem," FF 1940s, DB 1, Knapp Papers, ISU; *Milling and Baking News*, Dec. 25, 1973; Willard Cochrane to Orville Freeman, Feb. 15, 1963, FF 1963, Willard W. Cochrane memoranda (1), DB 10, 144.K.8 4, Freeman Papers, MHS; Voorhis, *American Cooperatives*, 90.

7. THE STATE AND AGRICULTURAL ORGANIZATION

1. Helmberger and Hoos, "Economic Theory of Bargaining," 1278; Gordon, *New Deals*, 10; John Mark Hansen, *Gaining Access: Congress and the Farm Lobby, 1919–1981* (Chicago: University of Chicago Press, 1991), 88; McGovern quoted in *City East: A Magazine for New Yorkers* 2 (Oct. 1968): 25.

2. Richard S. Kirkendall, *Social Scientists and Farm Politics in the Age of Roosevelt* (Columbia: University of Missouri Press, 1966), 18, 20, 24, 27, 40–41, 46–49; Hansen, *Gaining Access*, 71; Theda Skocpol and Kenneth Finegold, "State Capacity and Economic Intervention in the Early New Deal," *Political Science Quarterly* 97 (summer 1982): 260.

3. Kirkendall, *Social Scientists and Farm Politics*, 66–67; Theodore Saloutos, *The American Farmer and the New Deal* (Ames: Iowa State University Press, 1982), 88; Skocpol and Finegold, "State Capacity and Economic Intervention," 268–69; Murray R. Benedict, *Farm Policies of the United States, 1790–1950: A Study of Their Origins and Development* (New York: Twentieth Century Fund, 1953), 285, 314, 348.

4. Skocpol and Finegold, "State Capacity and Economic Intervention," 270–75; Luther G. Tweeten, *Foundations of Farm Policy*, 2d ed. (Lincoln: University of Nebraska Press, 1979), 460; Benedict, *Farm Policies of the United States*, 389, and *Can We Solve the Farm Problem*, 257–58. Soybean prices were supported every year since 1941 except 1975. Because their production was not controlled, soybeans could be grown on land idled by the corn and wheat program. *Soybeans: Background for 1985 Farm Legislation*, Sept. 1984, Agriculture Information Bulletin no. 471, Economic Research Service, USDA, 15.

5. Allen J. Matusow, *Farm Policy and the Truman Years* (Cambridge MA: Harvard University Press, 1967), 174, 181, 186–87.

6. Byron C. Hulsey, "'Back on the Beam of Freedom': Everett Dirksen's Election to the Senate and the Politics of 1950," paper delivered to the Mid-America History Conference, Sept. 13, 1996, 2, 6; Matusow, *Farm Policy and the Truman Years*, 178, 196–98, 200, 217–18, 220–21; Cochrane and Ryan, *American Farm Policy*, 73–74; Hansen, *Gaining Access*, 124–25.

7. Matusow, *Farm Policy and the Truman Years*, 247–48; Cochrane and Ryan,

American Farm Policy, 32–33; Schapsmeier and Schapsmeier, *Ezra Taft Benson*, 88; Ezra Taft Benson, *Freedom to Farm* (Garden City NY: Doubleday, 1960), 23.

8. Roy F. Hendrickson to the GTA Annual Convention, Dec. 15, 1954, FF 56, DB 15, M63-014, Cooperative League of USA Papers, WHS; Schapsmeier and Schapsmeier, *Ezra Taft Benson*, 78, 92, 94.

9. Schapsmeier and Schapsmeier, *Ezra Taft Benson*, 6–7; Ezra Taft Benson, *Crossfire: The Eight Years with Eisenhower* (Garden City NY: Doubleday, 1962); Bruce E. Field, "No Monolith Here: U.S. Farm Organizations in the First Decade of the Cold War," paper presented to the Center for Recent United States History, University of Iowa, Apr. 20, 1996, 25–28; George S. McGovern, *Grassroots: The Autobiography of George McGovern* (New York: Random House, 1977), 52; R. O. Meyer to McGovern, Oct. 20, 1961, DB Senate Files (Dec. 1980), 1960 campaign, 1962 more, McGovern Papers, MLPU.

10. *A Graphic Summary of South Dakota*, Census Data Center, Agricultural Experiment Station, Department of Rural Sociology, SDSU, 1993, 28; Robert Nelson to Melvin Hovland, Nov. 25, 1955, FF 1955 Re Nelson, Robert, and Robert Nelson to John Baker, n.d., FF 1956 Re Nelson, Robert, both in DB 1954–58 Selected Correspondence and Related Materials, McGovern Papers, MLPU; Herron Runestad to Eisenhower, Mar. 23, 1953, DB 329 GF 18-A Benson: Pro, FF 18-A Benson: Con, Eisenhower Library; Testimony of South Dakota Farmers Union, State Political Party Platform Hearings, 1972, FF Gen Farm Platform and DCD reorganization, DB 0056, Abourezk Papers, USD; Robert G. Lewis to Democratic and Farmers for Kennedy-Johnson Leaders, Oct. 22, 1960, FF 2, DB 5, Iowa Institute of Cooperation Papers, ISU.

11. McGovern speech to North Dakota Democratic Convention, FF Speech May 12, 1954, DB 1954, Speeches, Statements, McGovern Papers, MLPU.

12. Scott Heidepreim, *A Fair Chance for a Free People: A Biography of Karl E. Mundt, United States Senator* (Madison SD: Leader, 1988), 135; Schapsmeier and Schapsmeier, *Ezra Taft Benson*, 156, 192–93, 241; Wisconsin Agriculturalist Poll, released Mar. 1, 1956, FF 1958, DB 327 GF 18-A, Eisenhower Library; Anderson, "Mission, History, and Times, 138.

13. McGovern to Robert Nelson, n.d., Nelson to McGovern, July 5, 1956, McGovern to Nelson, July 9, 1956, Roy Glover to Robert Nelson, Nov. 19, 1956, all in FF 1956 Re Nelson, Robert, DB 1954–58 Selected Correspondence and Related Materials, McGovern Papers, MLPU; Herbert T. Hoover, "Farmers Fight Back: A Survey of Rural Political Organizations, 1873–1983," *South Dakota History* 13 (spring–summer 1983), 139.

14. See box 329, GF 18-A Benson: Pro at the Eisenhower Library for many NFO letters seeking Benson's removal. The South Dakota counties are found in FF 13/3,

DB 13, NFO Papers, ISU; videocassette TV-10 1-hour news, FF 25, DB 15, and NFO press release, FF 8, DB 12, both in NFO Papers, ISU; *Adams County Free Press*, Sept. 22, 1955; Iowa NFO BOD to all Iowa NFO County Officers, Dec. 24, 1957, FF 13, DB 7, Oren Lee Staley to James Patton, Dec. 11, 1956, James Patton to Oren Lee Staley, Dec. 17, 1956, FF 13, DB 7, all in ser. 4, NFU Papers, UCB; Schapsmeier and Schapsmeier, *Ezra Taft Benson*, 177; R. V. Fitzgerald Jr. to Eisenhower, Jan. 4, 1957, DB 329 GF 18-A Benson: Pro, FF 18-A Benson: Con, Eisenhower Library. George McGovern's Alpha File indicates that McGovern sent a telegram to Eisenhower on Apr. 20, 1957, and asked for Benson's removal, but the actual telegram could not be found.

15. Francis Case Reports from the U.S. Senate, "The 'Benson' Letter to Farmers," Jan. 30, 1957, and L. R. Houck to Francis Case, Feb. 14, 1957, FF 1958, DB 327 GF 18-A, Eisenhower Library; *Rapid City Journal*, Feb. 14, 1957; *Daily Plainsman*, Oct. 14, 1957; Reverend Luther O'Brien to Eisenhower, Oct. 19, 1957, DB 329 GF 18-A Benson: Pro, FF 18-A Benson: Con, Eisenhower Library; Heidepriem, *Fair Chance for a Free People*, 225; *Sioux City Journal*, Nov. 14, 1957; *Daily Plainsman*, Oct. 13, 1957.

16. McGovern speech, "The Brannan Plan," Mar. 5, 1958, and McGovern to Charles Brannan, Mar. 27, 1958, DB 1958 Correspondence Re legislation, speeches, remarks 1/box, FF Feb. 25, 1958, McGovern Papers, MLPU; Reo M. Christenson, *The Brannan Plan: Farm Politics and Policy* (Ann Arbor: University of Michigan Press, 1959), 169; *Brookings Register* clipping, 1960, FF Senate Races in South Dakota, DB 4989, 329-87-0099-428, McGovern Research 2/6, James Abdnor Papers, SDSHS.

17. Joe Foss with Donna Wild Foss, *A Proud American: The Autobiography of Joe Foss* (New York: Simon & Schuster, 1992), 217, 226–31; Heidepriem, *Fair Chance for a Free People*, 225; *Daily Republic*, Oct. 25, 1958; "Foss and the Farmer," FF 1958 Farm Issue, DB 1958 File Campaign and other items, McGovern Papers, MLPU; Karl Mundt to Hugh Agor, Nov. 19, 1958, FF 1, DB 38, RG 1, Karl Mundt Papers, DSU; *Sioux Falls Argus Leader*, Aug. 27, 1958; *Chicago Sun-Times*, Apr. 26, 1958.

18. William V. Shannon, "Into the Badlands," *New York Post*, Dec. 30, 1959; *Washington Post*, June 10, 1960; "King Karl Should Be Retired," FF 1960 Re Mundt, DB 1959–61 Selected Correspondence and Related Materials, MLPU; *Daily Republic*, Jan. 20, 1960; "Senator Karl E. Mundt's Work in Behalf of Agriculture and the Farmer" and unidentified newspaper clipping, scrapbook no. 99, Mundt Papers, DSU; *Salem Special*, Jan. 28, 1960.

19. *Lake Preston Times* article, n.d., and press release, Director of Food for Peace, FF articles 1961, DB May–June 1959, Speeches, Statements, Remarks, box

3, McGovern Papers, MLPU; "Survey of Political Attitudes: South Dakota, July 1962," DB 1962 Campaign, MLPU.

20. Don F. Hadwiger and Ross B. Talbot, *Pressures and Protests: The Kennedy Farm Program and the Wheat Referendum of 1963* (Ames: Iowa State University Press, 1965), 49.

21. Cochrane and Ryan, *American Farm Policy*, 36–38, 40; Hadwiger and Talbot, *Pressures and Protests*, 25; Willard Cochrane, "Some Observations of an Ex-Economic Advisor, or What I Learned in Washington," FF 1964 Willard Cochrane Memoranda (1), DB 11, 144.K.8.5(B), and John F. Kennedy, "Agricultural Policy for the New Frontier," FF 1961 Willard Cochrane Memoranda, DB 4, 144.K.8.3, Freeman Papers, MHS.

22. Cochrane and Ryan, *American Farm Policy*, 42; Hadwiger and Talbot, *Pressures and Protests*, 203, 214, 243; Hansen, *Gaining Access*, 153; Willard Cochrane to Orville Freeman, June 16, 1964, FF 1964 Willard Cochrane Memoranda (1), DB 11, 144.K.8.5(B), Freeman Papers, MHS.

23. Willard Cochrane to Orville Freeman, June 5, 1964, FF W. W. Cochrane memoranda (1), DB 11, 144.K.8.5(B), Freeman Papers, MHS.

24. Hansen, *Gaining Access*, 153–54; Hadwiger and Talbot, *Pressures and Protests*, 319–20, 324; Cochrane and Ryan, *American Farm Policy*, 47; Willard Cochrane to Orville Freeman, June 17, 1963, FF 1963, Willard Cochrane memoranda (1), DB 10, 144.K.8.4, Freeman Papers, MHS; Willard W. Cochrane and C. Ford Runge, *Reforming Farm Policy: Toward a National Agenda* (Ames: Iowa State University Press, 1992), 48–50. The 1970 law also limited total payments to farmers to fifty-five thousand dollars and introduced the generic "set-aside," as opposed to a specific restriction on growing a particular crop. The set-aside provision allowed farmers greater flexibility to plant nonquota crops. The 1973 law repealed the certificate program and replaced it with the target price system. *Wheat: Background for the 1985 Farm Legislation*, USDA, Economic Research Service, Agriculture Information Bulletin no. 467, 19.

25. Robert G. Lewis, "Competition and Cooperation in the Pricing of U.S. Wheats in Export Markets," Third National Wheat Utilization Research Conference, Kansas State University, Nov. 1964, FF S. 1946, carton 45, McGovern Papers, MLPU; Cochrane and Ryan, *American Farm Policy*, 39, 43; "NFO Farm Program Statement," c. 1955, FF 3, DB 12, NFO Papers, ISU.

26. Willard W. Cochrane, *Farm Prices: Myth and Reality* (Minneapolis: University of Minnesota Press, 1958), 167; Bruce Gardner, *The Governing of Agriculture* (Lawrence KS: International Center for Economic Policy Studies, 1981), 49; Willard Cochrane to Orville Freeman, Feb. 15, 1963, FF 1963, Willard W. Cochrane memoranda (1), DB 10, 144.K.8 4, Freeman Papers, MHS; James Patton speech to

the National Farm Institute, Feb. 18, 1956, FF 9, DB 24, Patton Papers, UCB; *Aberdeen American News*, June 2, 1960.

27. Quoted in GTA *Digest*, June–July 1963.

28. Thatcher speech to Farmers Union rally in Pipestone MN.

29. Hansen, *Gaining Access*, 90; Lynitta Aldridge Sommer, "Illinois Farmers in Revolt: The Corn Belt Liberty League," *Illinois Historical Review* 88 (winter 1995): 223, 228; Lane quote from John E. Miller, "Rose Wilder Lane and Thomas Hart Benton: A Turn toward History during the 1930s," *American Studies* 37 (fall 1996): 88; James Holt, "The New Deal and the American Anti-Statist Tradition," in John Braeman, Robert H. Bremner, and David Brody, *The New Deal: The National Level*, vol. 1 (Columbus: Ohio State University Press, 1975), 27.

30. James N. Giglio, "New Frontier Agricultural Policy: The Commodity Side, 1961–1963," *Agricultural History* 61 (summer 1987): 67; C. N. Hiebert to Frank Carlson, Feb. 19, 1962, FF General Correspondence, DB 231, Carlson Papers, KSHS; Hansen, *Gaining Access*, 130–36.

31. Quoted in Hadwiger and Talbot, *Pressures and Protests*, 190.

32. Dean, "Farm Policy Debate," 33; Cochrane and Ryan, *American Farm Policy*, 30.

33. Dean Hamilton to Frank Carlson, Sept. 23, 1962, and Ernest F. Quick to Frank Carlson, June 5, 1962, FF General Correspondence, DB 231, Carlson Papers, KSHS; Giglio, "New Frontier Agricultural Policy," 61, 65; Hansen, *Gaining Access*, 148, 151, 153; Cochrane and Ryan, *American Farm Policy*, 39, 43–44. Dole was forced to run against Breeding because of redistricting.

34. *Farm Quarterly*, fall 1960, quoted in *Wall Street Journal*, Mar. 6, 1961; Cochrane, "Some Observations"; Cochrane and Ryan, *American Farm Policy*, 46.

35. "Save Family Farm through Free Market Policy, Lugar Says," press release, Sept. 14, 1974, FF Republican Correspondence: Sept. 1974, DB BH, Butz Papers, PU; Friedberger, *Shake-Out*, 9; William P. Browne, "Challenging Industrialization: The Rekindling of Agrarian Protest in a Modern Agriculture, 1977–1987," *Studies in American Political Development* 7 (spring 1993): 16; Sommer, "Illinois Farmers in Revolt," 225; *Dakota Farmer*, May 1972, 10; Bruce Gardner, "The Federal Government in Farm Commodity Markets: The Recent Reform Efforts in a Long-Term Context," *Agricultural History* 70 (spring 1996): 194.

36. Helen C. Farnsworth, "Imbalance in the World Wheat Economy," *Journal of Political Economy* 66 (Feb. 1958): 9–10; Giglio, "New Frontier Agricultural Policy," 58, 67; Dale E. Hathaway, *Government and Agriculture: Public Policy in a Democratic Society* (New York: Macmillan, 1963), 298; John Kenneth Galbraith, "Economic Preconceptions and the Farm Policy," *American Economic Review* 44 (Mar. 1954): 52; Gerald A. Harrison and Earl O. Heady, "Acreage Diversion Re-

sponse under the 1961–1970 Feed Grain Program," Center for Agricultural and Rural Development Report 71 (Mar. 1977): 14; Alan R. Bird, *Surplus: The Riddle of American Agriculture* (New York: Springer, 1962), 35; Fite, *American Farmers*, 110–12.

37. Hansen, *Gaining Access*, 26–77.

38. Mr. and Mrs. Rudolph Kloehn and family to Karl Rolvaag, Sept. 16, 1961, FF Favorable NFO, DB 110.F.17.14(F), and Russel Schwandt to Karl Rolvaag, Oct. 29, 1964, DB 110.F.18.1(B), Rolvaag Papers, MHS; Iowa NFO Board of Directors to All County Officers, Dec. 24, 1957, FF 13, DB 7, ser. 4, NFU Papers, UCB; Patton, *Case for Farmers*, 48; Dechant speech to the National Association of Wheat Growers, Jan. 11, 1972, FF 41, DB 3, ser. 2, NFU Papers, UCB; Staley quoted in Frank O. Leuthold, "Agrarianism and Farm Organizations in the United States," in *The Agrarian Tradition in American Society: A Focus on the People and the Land in an Era of Changing Values*, A Bicentennial Forum, The Institute of Agriculture, The University of Tennessee, June 1976, 97.

39. Mait M. Holt to Bob Dole, Mar. 20, 1962, FF General Correspondence, DB 231, Carlson Papers, KSHS; Hansen, *Gaining Access*, 158, 174, 176.

40. Fite, *American Farmers*; Hansen, *Gaining Access*, 165, 167–69, 182–84.

41. Giglio, "New Frontier Agricultural Policy," 57; *Business Week*, Oct. 1, 1960; Cochrane, "Some Observations"; Cochrane and Ryan, *American Farm Policy*, 49, 59; Torgerson, *Producer Power at the Bargaining Table*, 23–24.

42. Hansen, *Gaining Access*, 187–91; Cochrane and Ryan, *American Farm Policy*, 51–52; *NFO on the Move*, Oct. 5, 1973; Thatcher speech to Farmers Union rally in Pipestone MN.

43. McGovern to Harold Mayer, Apr. 17, 1965, FF Press Release Apr. 19, 1965, Senator McGovern to Mayor Wagner, carton 3, McGovern Papers, MLPU; Matusow, *Farm Policies and Politics* and *Unraveling of America*.

44. Anthony J. Badger, *The New Deal: The Depression Years, 1933–1940* (New York: Hill & Wang, 1989), 118–19; Michael Kazin, *The Populist Persuasion: An American History* (New York: Basic Books, 1995), 137–38; Mark R. Finlay, "Dashed Expectations: The Iowa Progressive Party and the 1948 Election," William C. Pratt, "The Farmers Union and the 1948 Henry Wallace Campaign," and Warren, "The 'People's Century' in Iowa," all in *Annals of Iowa* 49 (summer 1988): 329–93.

45. Matusow, *Farm Policies and Politics*, 200, 209, 227; Dean, "The Farm Policy Debate," 41; Hansen, *Gaining Access*, 120, 166.

46. McGovern, statement to Congress, FF Writings 1953, DB 1953: Speeches, Statements: Re Legislation and other matters, McGovern Papers, MLPU; Galbraith, *American Capitalism*, 114–17; "Scholars Starting to Advise McGovern," *New York Times*, June 18, 1972.

47. Thomas Cosgrove and John Edelman (Textile Workers) to McGovern, Feb. 20, 1957, Kenneth Peterson (IUE) to McGovern, Apr. 4, 1957, and McGovern to Kenneth Peterson, Dec. 19, 1957, FF Speech: Economic Cost of Discrimination, DB 1957, Correspondence, Re Legislation, Speeches, Remarks, McGovern Papers, MLPU; Emil Loriks to McGovern, Mar. 3, 1958, FF Feb. 25, 1956, H.R. 10966, DB 1958, Correspondence Re Legislation, Speeches, Remarks, McGovern Papers, MLPU.

48. *In Fact* 12 (Feb. 18, 1946); Holton Davenport to Case, May 10, 1946, J. T. Saunders to Case, May 27, 1946, Ray Stoltz to Case, June 1, 1946, H. M. Truex to Case, Feb. 28, 1946, M. A. Loros to Senator Chan Gurney, May 23, 1946, Floyd Wilkerson to Senator Harlan Bushfield, May 15, 1946, all in FF South Dakota: Pro, file drawer 26, Case Papers, DWU; Arthur Moore, *The Farmer and the Rest of Us* (Boston: Little, Brown 1945), 113.

49. In 1956 McGovern received $3,000 from the AFL-CIO Committee on Political Education (COPE), $750 from the UAW, $150 from the International Lady Garment Workers, and $500 from the Machinists. Since the entire campaign only cost $14,000, over 30 percent of the money came from unions. "1956 Campaign Labor Contributions," FF 1958 Labor Issue, DB 1958 File Campaign and other items, McGovern Papers, MLPU; McGovern, *Grassroots*, 66; *Machinist*, July 24, 1958; Heidepriem, *Fair Chance for a Free People*, 227; memo, William E. Sweisgood to Senator Smathers, June 29, 1960, 1960 Congressional File, McGovern, LBJ Library.

50. *Chicago Sun-Times*, Apr. 26, 1958; "Information Furnished by Richard Schifter, Based on Confidential Discussions with Dr. William Farber, USD," "Foss and the Labor Issue," and McGovern and J. T. McMullen to Bob Hipple, Western Union Wire sent Oct. 23, 1958, all in FF 1958 Labor Issue, DB 1958 File Campaign and other items, McGovern Papers, MLPU; McGovern to Harald Cooley, Feb. 26, 1959, FF 1959 Re Agriculture: House Committee on Agriculture, DB 1959 Correspondence Re legislation: Agriculture Aa-Rz, McGovern Papers, MLPU.

51. Hansen, *Gaining Access*, 172, 203–4; "Senate Fails to Adopt Cloture on Mansfield Effort to Begin Debate on Bill to Repeal Right-to-Work Section 14b of Taft-Hartley Act," *1965 Congressional Quarterly Almanac* 21:1078; during the avalanche of legislation during the Johnson years Senator Mansfield sought to repeal Taft-Hartley. McGovern voted against invoking cloture (ending debate), one of only five northern Democrats to do so, so the measured died. Irving Bernstein, *Guns or Butter: The Presidency of Lyndon Johnson* (New York: Oxford University Press, 1996), 307–13; George McGovern to Ralph Massie, Aug. 8, 1965, General File, Ta4/CO303, LBJ Library; *Business Week*, July 15, 1972, 22.

52. Harry L. Graham letter, Apr. 13, 1970, FF 19, DB 10, ser. 3, NFU Papers, UCB;

Cochrane and Ryan, *American Farm Policy*, 27.

53. Cochrane and Ryan, *American Farm Policy*, 88–89; *Congressional Record*, 86th Cong., 1st sess. (Mar. 17, 1959), 105, pt. 4: A2287; Cochrane, "Some Observations"; Hansen, *Gaining Access*, 170, 186; Giglio, "New Frontier Agricultural Policy," 58, 67–70; Cochrane and Ryan, *American Farm Policy*, 40, 42, 49.

54. Herschel D. Newsom to the Fifth Annual Pacific Northwest Farm Forum, Feb. 10, 1958, FF 3, DB 12, NFO Papers, ISU; Cochrane and Ryan, *American Farm Policy*, 50; Lewis, "Competition and Cooperation"; Benedict, *Farm Policies of the United States*, 302.

55. Hansen, *Gaining Access*, 187, 84, 93–94.

CONCLUSION

1. Bureau of the Census, *1992 Census of Agriculture*, Geographic Area Series, vol. 1, part 51, *United States: Summary and State Data*, 4.

2. Hiram M. Drache, *Tomorrow's Harvest: Thoughts and Opinions of Successful Farmers* (Danville IL: Interstate, 1978), 200.

3. David Gilkerson to George McGovern, Mar. 1, 1958, FF Feb. 25, 1958 H.R. 10966, DB 1957, Correspondence, Re legislation, Speeches, Remarks, McGovern Papers, MLPU.

4. John R. Burk to Frank Carlson, Jan. 17, 1962, FF General Correspondence, DB 231, Carlson Papers, KSHS.

5. *Billings Gazette*, Apr. 3, 1975.

6. Willard Cochrane to Orville Freeman, June 16, 1964, and Cochrane, "Some Observations," FF 1964 Willard Cochrane Memoranda (1), DB 11, 144.K.8.5(B), Freeman Papers, MHS; Willard W. Cochrane, *The City Man's Guide to the Farm Problem* (Minneapolis: University of Minnesota Press, 1965), vii.

7. *Farm Journal*, Apr. 1959, poll included as part of FF Mar. 25, 1959, Statement: Subcommittee on Agricultural Conservation, DB Mar.–Apr. 1959 Speeches, Statements, etc. Re Legislation and other matters, box 2, McGovern Papers, MLPU.

8. Linda Hasselstrom, *Land Circle: Writings Collected from the Land* (Golden CO: Fulcrum, 1991), 69, 74; Unger, *Leaving the Land*, 12.

9. Hasselstrom, *Land Circle*, 73; Roosevelt speech "The Strenuous Life" delivered to the Hamilton Club, Chicago, Apr. 10, 1899, in Charles Hurd, *A Treasury of Great American Speeches* (New York: Hawthorne, 1959), 151–53; Lasch, *True and Only Heaven*, 79.

10. Hasselstrom, *Land Circle*, xix, 79.

11. Hasselstrom, *Land Circle*, 26–27.

12. Hasselstrom, *Land Circle*, xx, 63, 69, 77; Christopher Lasch, *True and Only Heaven*, 17 and passim; Lasch, *Revolt of the Elites*, 39. See also *The Culture of*

Narcissism: American Life in An Age of Diminishing Expectations (New York: Norton, 1979).

13. Garry Wills, *John Wayne's America: The Politics of Celebrity* (New York: Simon & Schuster, 1997), 303; Fite, *American Farmers*, 123.

14. Amassa Alden, "Don't Leave the Farm," reprinted in Robert Walker, "The Poet and the Rise of the City," *Mississippi Valley Historical Review* 49 (June 1962): 87.

15. Creedmore Fleenor quoted in Walker, "The Poet and the Rise of the City," 88.

16. Agnew quoted in Jonathan Reider, "The Rise of the 'Silent Majority,'" in Fraser and Gerstle, eds., *Rise and Fall of the New Deal Order*, 262; Fred Setterberg, *The Roads Taken: Travels through America's Literary Landscapes* (Athens: University of Georgia Press, 1993), 49; James R. Shortridge, *The Middle West: Its Meaning in American Culture* (Lawrence: University Press of Kansas, 1989), 130; Luke J. Schissel to John Culver, July 21, 1968, no FF, DB 44 Legislative Correspondence, 1968–69, Culver Papers, UI.

17. Edwin Markham, "The Man with the Hoe," quoted in Wilson Gee, *The Social Economics of Agriculture* (New York: Macmillan, 1942), 20.

18. C. Elizabeth Raymond, "The Creation of America's Rural Heartland: An Essay on Prairie Midwestern Regional Identity," 13, unpublished paper delivered to the Newberry Seminar, 1994; David Danbom, *Born in the Country: A History of Rural America* (Baltimore: Johns Hopkins University Press, 1995), 151; Shortridge, *Middle West*, 33; Jeffrey Meyers, *Edmund Wilson: A Biography* (Boston: Houghton Mifflin, 1995), 30; H. L. Mencken, *American Mercury*, Jan. 1931; quote from the Communist Manifesto reprinted in Jeffrey L. Pasley, "The Idiocy of Rural Life," *New Republic*, Dec. 8, 1986, 24.

19. D. N. McCloskey, "Bourgeois Virtue," *American Scholar* 63 (spring 1994): 177–91; David B. Danbom, "The Professors and the Plowmen in American History Today," *Wisconsin Magazine of History* 69 (winter 1985–86): 123; Lawrence Goodwyn, "The Cooperative Commonwealth and Other Abstractions: In Search of a Democratic Premise," *Marxist Perspectives* 3 (summer 1980): 14; Carl Ubbelohde, "History and the Midwest as a Region," *Wisconsin Magazine of History* 78 (autumn 1994): 44.

20. Joseph E. Baker, "The Midwestern Origins of America," *American Scholar* 17 (winter 1947–48): 58.

21. T. J. Jackson Lears, "The Concept of Cultural Hegemony: Problems and Prospects," *American Historical Review* 90 (June 1985): 567; John Patrick Diggins, "Comrades and Citizens: New Mythologies in American Historiography," *American Historical Review* 90 (June 1985): 616. Referring to early America, for

example, Paul Gilje notes that for the last twenty-five years "some scholars have argued that capitalism was not always a part of the American mentality." He edited the "Special Issue on Capitalism in the Early Republic," *Journal of the Early Republic* 16 (summer 1996): 159.

22. John E. Miller, *Laura Ingalls Wilder's Little Town: Where History and Literature Meet* (Lawrence: University of Kansas Press, 1994), 36; Unger, *Leaving the Land*, 9–10; directors Jeanne Jordan and Steven Ascher, *Troublesome Creek: A Midwestern* (Artistic License, 1995). This film won the Grand Jury Prize and the Audience Award at the 1996 Sundance Film Festival. Louis Hartz, *The Liberal Tradition in America: An Interpretation of American Political Thought since the Revolution* (New York: Harcourt, Brace, 1955); Richard Hofstadter, *The American Political Tradition and the Men Who Made It* (New York: Knopf, 1951 [1948]), viii. This is not to say that Hofstadter liked this political tradition, farmers, or small towns—he "went so far as to blame most of the violence in American history on small-town culture," for example. Daniel Joseph Singal, "Beyond Consensus: Richard Hofstadter and American Historiography," *American Historical Review* 89 (Oct. 1984): 991 n.29.

23. Wills, *John Wayne's America*, 11, 309, 311.

24. David Harlan, *The Degradation of American History* (Chicago: University of Chicago Press, 1997), 43, 51.

25. Mrs. E. C. Hallstein to Cliff Benson, Sept. 30, 1964, FF Favorable NFO, DB 110.F.17.14(F), Rolvaag Papers, MHS.

26. LeRoy Ashby, *William Jennings Bryan: Champion of Democracy* (Boston: Twayne, 1987), 30.

27. Wood, "The Enemy Is Us"; D. N. McCloskey, "The Open Fields of England: Rent, Risk, and the Rate of Interest, 1300–1815," in David Galenson, ed., *Markets in History: Economic Studies of the Past* (New York: Cambridge University Press, 1989), 5–51; Hartz, *Liberal Tradition in America*, 118. According to Brian Harding, the Garden of the World Henry Nash Smith describes "is a market garden and this bower of paradise is one in which the American spirit of business enterprise may produce the most luxuriant vegetation." Brian Harding, "The Myth of the Myth of the Garden," in Ian F. A. Bell and D. K. Adams, eds., *American Literary Landscapes: The Fiction and the Fact* (New York: St. Martin's, 1989), 58–59. Henry Nash Smith, *Virgin Land: The American West as Symbol and Myth* (Cambridge MA: Harvard University Press, 1950).

28. Daniel Bell, *The Cultural Contradictions of Capitalism* (New York: Basic Books, 1976), 10–12, 14; Bruchey quoted in David A. Horowitz, *Beyond Left and Right: Insurgency and the Establishment* (Urbana: University of Illinois Press, 1997), 5.

29. "The Hit Men," *Newsweek*, Feb. 26, 1996; "Time to Debunk the Myths," *Business Week*, June 16, 1997; *Wall Street Journal*, Feb. 26–28, Mar. 3, 4, 1997. Even William Safire has doubts about the recent banking mergers: "Our financial institutions can go global without going gaga. I've never knocked greed, but this spreadeagled 'universality' is getting out of hand" (*New York Times*, Apr. 16, 1998); Ron Chernow, "The Monopoly That Went Too Far: John D. Rockefeller's Arsenal of Anticompetitive Weapons Ultimately Made Standard Oil the Focal Point of Public Ire," *Business Week*, May 18, 1998; "Magnetic Mania: In This Merged, Merged World, Anything Goes," *New York Times*, June 26, 1998.

30. Peter G. Carstensen, "How to Assess the Impact of Antitrust on the American Economy: Examining History or Theorizing?" *Iowa Law Review* 74 (July 1989): 1175.

31. Carstensen, "How to Assess the Impact of Antitrust," 1179–80. See also Peter Passell's columns about the questionable economic benefits of many mergers: "Do Mergers Really Yield Big Benefits?" May 14, 1998, and "When Mega-Mergers are Mega-Busts," May 17, 1998, both in the *New York Times*; Benjamin Barber, "Big=Bad, Unless It Doesn't," *New York Times*, Apr. 16, 1998; William G. Shepard, "Dim Prospects: Effective Competition in Telecommunications, Railroads, and Electricity," *Antitrust Bulletin* 42 (spring 1997): 166 ("Moreover, mergers often harm the partners and their shareholders, by fostering errors and inefficiency. Yet they remain a popular device to retain or restore monopoly"); Deirdre McCloskey, *The Rhetoric of Economics* (Madison: University of Wisconsin Press, 1985) and *If You're So Smart*; David Millon, "The Sherman Act and the Balance of Power," *Southern California Law Review* 61 (July 1988): 1219.

32. Rudolph J. R. Peritz, "Some Realism about Economic Power in a Time of Sectorial Change," *Antitrust Law Journal* 66 (1997): 247, 251–52; Rudolph Peritz, "A Counter-History of Antitrust Law," *Duke Law Journal* (Apr. 1990): 263, 266. The free market advocate's admission is reprinted in Jeffrey Friedman, "What's Wrong with Libertarianism," *Critical Review* 11, no. 3 (summer 1997), citing Richard Cornuelle's article "New Work for Invisible Hands," *Times Literary Supplement*, Apr. 5, 1991; "Growth of Factory-Like Hog Farms Divides Rural Areas of the Midwest," *New York Times*, June 24, 1998.

33. Lasch, *Revolt of the Elites*, 85–88; Harlan, *Degradation of American History*, xxxii.

34. Mr. and Mrs. George Heibult, Sept. 1959, FF 1959 Re Agriculture: Egg and Poultry, DB 1959 Correspondence Re Legislation: Agriculture Aa-Rz, McGovern Papers, MLPU; Robert Ostergren, "European Settlement and Ethnicity Patterns on the Agricultural Frontiers of South Dakota," *South Dakota History* 13 (spring–summer 1983): 67–68, 76; G. D. Lillibridge, "Small-Town Boys: Growing Up in

Mitchell in the 1920s and 1930s," *South Dakota History* 25 (spring 1995): 7–8, 15, 20; *Congressional Record*, 88th Cong., 2d sess. (June 1, 1964), 110, pt. 9: 11912.

35. FF McG Fund Letters, DB 329-87-0099 431, McGovern Research 5/6, Abdnor Papers, SDSHS; Reider, "The Rise of the 'Silent Majority,'" 259; William H. Chafe, *The Unfinished Journey: America since World War II* (New York: Oxford University Press, 1986), 459. For McGovern's 1980 election Gloria Steinem sent out a "Dear Friend" fund-raising letter seeking funds to stop the "emergency" in South Dakota.

36. Shortridge, *Middle West*, 67, 71.

37. Renee Drury and Luther Tweeten, *Have Farmers Lost Their Uniqueness?* Anderson Report ESO 2237, Department of Agricultural Economics, Ohio State University, 1995, i.

38. Garrison Keillor, "Sweet Home, Minnesota," *Time*, Mar. 24, 1997, 108.

39. Richard Critchfield, *The Villagers: Changed Values, Altered Lives: The Closing of the Urban-Rural Gap* (New York: Anchor, 1994), 435; Jean Bethke Elshtain, *Democracy on Trial* (New York: Basic Books, 1995), xv.

EPILOGUE

1. Thomas F. Stokes, Testimony to Senate Agriculture Committee, Jan. 26, 1999, 3; A Resolution Calling upon the U.S. Dept. of Justice and the Attorneys General of Minnesota, South Dakota, Iowa, and Nebraska to Investigate Collusive Practices in the Midwestern Food Processing Industries, Four-State Farm Price Crisis Forum, Sioux City IA, Jan. 30, 1999; "Agriculture: In the Mill," *Economist*, Mar. 20, 1999, 64; Ed Maixner, "Ag Sector Concentration Gets More Federal Attention", *Feedstuffs*, Jan. 25, 1999; William Heffernan, "Consolidation in the Food and Agriculture System," Report to the National Farmers Union, Feb. 5, 1999; Leland Swenson, "Merger Mania Troublesome for Family Producers," *National Farmers Union News*, Jan. 1999, 2 (mother quote); Senators Byron Dorgan and Chuck Hagel to President Clinton, Feb. 19, 1999. The Farmers Union has asked that Congress "establish a percentage of concentration that automatically triggers anti-trust action." Leland Swenson, testimony to the Senate Agriculture Committee, Jan. 26, 1999; press release, Senator Tom Daschle, "Daschle Tells Senate Agriculture Committee That Growing Industry Concentration Is Contributing to Farm Crisis; Presses Committee to Take Action to Halt Dramatic Loss of Producers," Jan. 26, 1999.

2. C. Robert Taylor, "Economic Concentration in Agribusiness," testimony to the United States Senate Committee on Agriculture, Nutrition, and Forestry, Jan. 22, 1999, 2; Jean Kinsey, *Concentration of Ownership in Food Retailing: A Review of the Evidence About Consumer Impact*, Working Paper 98-04, Retail Food

Industry Center, University of Minnesota, 13. The review concluded, however, that studies of the relationship between concentration and profitability produced "mixed evidence" and that "although research related to the question of monopoly power in the food industry has produced some intellectually interesting theories and measures of profitability, it has not richly informed consumers or public policy makers about the state of the industry as it operates today" (21–22). Dorgan and Hagel letter; Stokes testimony.

3. Roger D. Blair and Jeffrey L. Harrison, "Antitrust Policy and Monopsony," *Cornell Law Review* 76 (1991): 297–98; James Murphy Dowd, "Oligopsony Power: Antitrust Injury and Collusive Buyer Practices in Input Markets," *Boston University Law Review* 76 (1996): 1078–9; *U.S. v. Rice Growers Association of California*, 1986 WL 12562, 4 (E.D. Cal. 1986).

4. Mary Lou Steptoe, "The New Merger Guidelines: Have They Changed the Rules of the Game?" *Antitrust Law Journal* 61 (1993): 493–4; *U.S. v. Country Lake Foods, Inc.*, 754 F. Supp. 669 (D. Minn. 1990); Luciano Venturini, *Countervailing Power and Antitrust Policy in the Food System*, Sixth Joint Conference on Food, Agriculture, and the Environment, hosted by Center for International Food and Agricultural Policy, University of Minnesota, Aug.–Sept. 1998.

5. *U.S. v. United Tote, Inc.*, 768 F. Supp. 1064 (D. Del. 1991). The totalisator system manages betting at horse tracks. *FTC v. Cardinal Health, Inc.*, 12 F. Supp. 2d 34 (D.C. Cir. 1998).

6. *Eastman Kodak Co. v. Image Technical Services*, 504 U.S. 451 (1992); Michael S. Jacobs, "Market Power through Imperfect Information: The Staggering Implications of *Eastman Kodak Co. v. Image Technical Services* and a Modest Proposal for Limiting Them," *Maryland Law Review* 52 (1993): 336; Mark R. Patterson, "Product Definition, Product Information, and Market Power: Kodak in Perspective," *North Carolina Law Review* 73 (1994): 187; George Stigler, *The Organization of Industry* (Chicago: University of Chicago Press, 1968), 171–190; Market Access, 1995 Survey Results (Iowa Pork Producers Association, in cooperation with Iowa State University), 3; Alan Schwartz and Louis L. Wilde, "Intervening in Markets on the Basis of Imperfect Information: A Legal and Economic Analysis," *Cornell Law Review* 127 (1979): 667.

7. Michael S. Jacobs, "The New Sophistication in Antitrust," *Minnesota Law Review* 79 (1994): 53 n. 149, noting that the term "post-Chicago" "apparently originated" in Herbert Hovenkamp, "Antitrust Policy after Chicago," *Michigan Law Review* 84 (1985): 213; Jonathan B. Baker, "Recent Developments in Economics that Challenge Chicago School Views," *Antitrust Law Journal* 58 (1989): 645; Martin Shubik, "Game Theory, Law, and the Concept of Competition," *University of Cincinnati Law Review* 60 (1991): 285; Michael O. Wise, "Antitrust's

Newest 'New Learning' Returns the Law to Its Roots: Chaos and Adaptation as New Metaphors for Competition Policy," *Antitrust Bulletin* 40 (1995): 723–24; Michael S. Jacobs, "An Essay on the Normative Foundations of Antitrust Economics," *North Carolina Law Review* 74 (1995): 226; Lawrence A. Sullivan, "Post-Chicago Economics: Economists, Lawyers, Judges, and Officials in a Less Determinate Theoretical World," *Antitrust Law Journal* 63 (1995): 670.

8. "The correct rule of interpretation is, that if divers statutes relate to the same thing, they ought all to be taken into consideration in construing any one of them, and it is an established rule of law, that all acts in pari materia are to be taken together, as if they were one law." *United States v. Freeman*, 44 U.S. 556, 564 (1845); Robert Pitofsky, "The Political Content of Antitrust," *University of Pennsylvania Law Review* 127 (1979): 1065; William N. Eskridge Jr. and Philip P. Frickey, "Statutory Interpretation as Practical Reasoning," *Stanford Law Review* 42 (1990): 356.

INDEX

AAA. *See* Agricultural Adjustment Act
Abourezk, James: on rural outmigration, 8; on corporate farming, 22, 24
ADM. *See* Archer Daniels Midland
Agnew, Spiro, 167
Agribusiness Accountability Project, 21, 24, 30, 35
Agricultural Adjustment Act (AAA), 64, 65–66, 138
Agricultural History Society, 21
Alcoa, 15
AMPI (Associated Milk Producers Inc.), 132
American Farm Bureau Federation, 35, 95
Andreas, Dwayne, 127, 130
Andrews, Mark, 134
antitrust laws: conservative case for enforcement of, 17; and economic decentralization, 14–15; and farmer bargaining power, 177–81; non-economic considerations of, 15–17; reformulation of, 177–81
Appleby, Joyce, 5
Archer-Daniels-Midland (ADM), 12, 127–28, 172
Arnold, Thurman, 134
Ashby, LeRoy, 171
Associated Milk Producers Inc. (AMPI), 132
Ayres, Homer, 1

Bagley, Hughes, 53
Bain, Joe, 1; *Industrial Organization*, 10
Banks for Cooperatives, 114–16
Bell, Daniel, 172
Bender, Thomas, xi
Benson, Ezra Taft, 140–43
Bergland, Bob, 36, 121, 132, 134

Berry, Wendell: *The Unsettling of America*, 27
Bland, Richard, 43
Blobaum, Roger, 29
Bok, Derek, 12
Brannan, Charles, 139
Breeding, Floyd, 152
Breimyer, Harold, 9, 30–31, 94
Bryan, William Jennings, 14
Buchanan, Patrick, 172
Butz, Earl, 24, 27, 31, 65, 77, 79, 88, 133
Byrd, Harry, 73

Campbell, J. Phil, 35
CAP. *See* Common Agricultural Policy
Capper, Arthur, 67, 116, 153
Capper-Volstead Act, 88, 100, 106, 108, 111, 132, 133–34
captive supplies of livestock, 55
Cargill, 62; and international trade negotiations, 71; and meatpacking, 53; and multimarket concentration, 60; and Soviet grain sales, 79, 83
Carstenson, Peter, 173
Carter, Jimmy, 24, 78, 80, 132
Catholic Church: and corporate farming debate, 26–27; and NFO, 108
CCA (Consumers Cooperative Association), 116, 124
Celler-Kefauver Antimerger Act, 12
Center for Rural Affairs, 27–28, 34
Chernow, Ron, 173
Chicago school economics, 16
Chicago school of antitrust, 10, 15; weaknesses of, 11–13
chicken war, 72–74

256 INDEX

Christgau, Victor, 137
Church, Frank, 77
Clark, Richard, 44, 62, 104
Clodius, Robert, 89, 146
Cochrane, Willard, 74, 135, 146, 152, 156
Colorado, 44
commercialization of farming, 4
Committee on Economic Development, 88
Common Agricultural Policy (CAP): and GATT negotiations, 71–75; origins of, 71
concentration in food sector: and conglomerate growth, 41–42; post–World War II growth in, 40
Consumers Cooperative Association (CCA), 116, 124
Cooley, Harold, 155
Coolidge, Calvin, 64
cooperative organizations: and antitrust laws, 133–35; and Banks for Cooperatives, 114–16; and Capper-Volstead Act, 111–12; and corporate resistance, 116–17, 132–33; and defense against legislative attacks, 117–18; and exporting, 123; and food manufacturing, 122; and Hoover's farm program, 112–14; and mergers, 130, 135; origins, 109–11; political power, 118–22; and Sapiro, 112; value-added, 130–31. *See also* Consumers Cooperative Association; Far-Mar-Co; Farmland Industries; Grain Terminal Association; National Council of Farmer Cooperatives; National Farmers Organization; National Farmers Union
corn industry, 58–59
corporate farming: in American South, 28–29; and California agriculture, 29–30; and Catholic Church, 26; and criticism of big business, 22–24; early attempts at, 20; and land grant colleges, 25; and land reform, 30–31; laws restricting, 20–21, 36–37; and legislation, 32–36; and news media, 28; origins of, 20; and politics, 24–26; and republicanism, 31; social concerns about, 21–22
Council of Economic Advisors, 15
countervailing power: difficulties of, 13–14; as new dimension in antitrust analysis, 13–15
Cowden, Howard, 116
Critchfield, Richard, 176
Curtis, Carl, 46

Danbom, David, 168
Dewey, Donald, 16
Diggins, John Patrick, 169
Dirksen, Everett, 123
Dole, Robert, 14, 75, 78, 152, 154
Douglas, William, 13
Dulles, John Foster, 69

EEC. *See* European Economic Community
Eisenhower, Dwight, 3, 69, 84, 88, 119, 140
Eliot, T. S., xii
Ellender, Allen, 99
Elshtain, Jean Bethke, 176
environmentalists: concerns about corporate farming of, 27; and harm from agriculture, 6–7
European Economic Community (EEC): ban imports of hormone-treated beef, 48; and "chicken war," 72–74; as sign of rising protectionism, 71–74
Export Enhancement Program, 75

farm crisis of 1980s, ix; and new farm organizations, 3
Far-Mar-Co, 14, 127–29
farmer bargaining power, 13
Farmers Union Marketing and Processing Association, 123
Farmland Industries, 124–25
farm program: Benson's modifications to, 140–41; Brannan plan for, 139; and consumers, 155–57; and Eisenhower, 140–42; and federal budget, 160; and international trade, 161; and Kennedy, 145–47; and labor, 157–60; and operation of AAA, 138; origins of, 137–38; and political coalitions, 153–55; resistance to controls of, 148–52; unpopularity of Benson's, 141–43; and yield increases, 152–53
farmworkers, 29–30
Federal Trade Commission (FTC), 10, 11, 12, 14, 39, 43, 44, 60, 97, 133
Fite, Gilbert, 124, 155
Foley, Tom, 154
Food Action Committee, 22
food prices, 39–40, 155–56
food processing: and concentration levels, 40–41; and conglomerate mergers, 41–42
Ford, Gerald, 78, 104
Foss, Joe, 144, 159
Freeman, Orville: and bargaining legislation, 97; and farm program, 145; and

INDEX 257

rural outmigration, 6; and study of corporate farming, 32–33
Friedberger, Mark, 9
FTC. *See* Federal Trade Commission

Galbraith, John Kenneth: *American Capitalism*, 13; on countervailing power, 13, 89; on farm program, 153; on oligopoly, 1
GATT. *See* General Agreement on Tariffs and Trade
General Agreement on Tariffs and Trade (GATT), 66, 71, 74; and wheat dispute, 74–75
General Mills, 12
Genovese, Elizabeth-Fox, 17
Genovese, Eugene, 17
Genscher, Hans Dietrich, 74
Giglio, James, 155
grain belt: defined, 2–3
Grain Terminal Association (GTA): as success story, 14; growth of, 126; and ADM, 127; origins of, 125
Gramsci, Antonio, 169
Grange, 3, 26, 36, 161
Greenspan, Alan, 15–16
Griswold, A. Whitney, 16
GTA. *See* Grain Terminal Association

Hadwiger, Don, 146
Hamilton, David, 112
Hand, Learned, 15
Hansen, John, 161
Hathaway, Dale, 152
Hardin, Clifford, 24, 161
Harkin, Tom, 3
Harlan, David, 170–71, 174
Harris, Fred, 24, 31
Hart, Phil, 23
Hartz, Louis, 170
Harvey, Paul, 104
Hasselstrom, Linda, 165–67
Hawley, Ellis, 6
Helmberger, Peter, 89
Hickenlooper, Bourke, 101, 104
Hightower, Jim, 21, 25
Hills, Carla, 75
historians: and agriculture, x–xi; and farmer cooperatives, 109–10; and farmer libertarianism, 164; and monopoly problem, 5; New Rural, xi; and NFO, 85; and political economy, 17–18

Hoffman, A. C., 40
Hofstadter, Richard, 17
Holman, Currier, 53
Hoos, Sidney, 89
Hoover, Herbert: attempts to stabilize global commodity prices, 63–64; and cooperatives, 111; and domestic agricultural reforms, 64–65; farm program, 112–14
Hull, Cordell, 65
Humphrey, Hubert, 7, 25, 108, 118–19, 152
Hurd, Douglas, 74

IBP (Iowa Beef Processors), 51–53
Ickes, Harold, 2
international agricultural trade: and CAP, 71–73; and chicken war, 72–74; and diplomatic obstacles, 74–75; and market entry, 80–83
International Trade Organization, 66
International Wheat Agreements (IWA): origins of, 65; postwar, 66–69
Iowa, 44, 51, 54–55
Iowa Beef Processors (IBP), 51–53
ITO, 66
ITT, 15; as corporate farmer, 31; and toppling of Salvador Allende, 23
IWA. *See* International Wheat Agreements

Jackson, Henry, 79, 117
Jackson, Robert, 15
Javits, Jacob, 156
Jefferson, Thomas, 16, 32
Johnson, D. Gale, 75
Johnson, Lyndon Baines, 85
John XXIII, Pope, 27

Kansas: and corporate farming, 21
Kefauver, Estes, 121
Kellog, 12
Kennedy, John F.: farm program, 145–47
Kennedy, Robert, 99
Kennedy, Ted, 24
Kissinger, Henry, 65, 77, 79
Knowlton, Dick, 51
Knudson, Harold, 118

LaGuardia, Fiorello, 155
Lanzillotti, Robert, 10
Lasch, Christopher, x, 18
Lincoln, Abraham, xi; defining the grain belt, 2

258 INDEX

Loveland, Albert, 139
Lucas, Scott, 139
Lugar, Richard, 152
Lynch, David, 6

Matusow, Allen: *The Unraveling of America*, 157
Marx, Karl, 2
McCloskey, Deirdre, 16–17
McGovern, George: and agricultural politics of 1950s, 140–45; and corporate farming, 25–26, 31, 35; farm policy proposals of, 143–44, 147–48; and Kennedy farm program, 145; and labor, 158–60; and monopolies, 39; and rural outmigration, 8; and Soviet grain sales, 75–76
McNary-Haugen bill, 64
McNicholls, James: *A Theoretical Analysis of Imperfect Competition with Special Application to Agricultural Industries*, 39
Meat Import Act of 1964, 47, 72
meatpacking: and aggregate demand, 46–47; and changes in market structure, 50–51; and farmer bargaining, 48; and feeding sector, 43–44; and imports, 47–48; and labor costs, 54–56; mergers in, 50; and NFO, 48; origins of antitrust concern about, 42–43; and profits, 48–50; reconcentration of, 56–57; and retail and restaurant sector, 44–46; technological changes in, 54. *See also* Cargill; Iowa Beef Processors
Melcher, John, 161
Mencken, H. L., 168
mergers, 5, 172–74; in meatpacking, 50
Mezvinsky, Ed, 132
Miller, John, 170
Minnesota, 56
Mondale, Theodore, 1
Mondale, Walter, 70; and bargaining legislation, 98, 126; and cooperatives, 132
monopoly capitalism, 1
monopoly problem, 1; defined, 3–4; nature of, 3–4, 6; new measures of, 9–11; origins of, 4–5; social consequences of, 2, 6–8
Morse, Wayne, 23, 73
Mueller, Willard, 10, 13
Mundt, Karl, 25, 75, 99, 142–45, 159

Nader, Ralph, 28, 29
National Commission for the Review of the Antitrust Laws, 133
National Commission on Food Marketing, 50, 59, 98, 122
National Council of Farmer Cooperatives (NCFC), 109, 117, 119, 131
National Farmers Organization (NFO): and bargaining legislation, 95–101; and collective bargaining program, 87–89; and corporate farming, 21–22, 34; early legislative program of, 86–87; extent of operations of, 94–95; financial problems of, 106; and historians, xi; holding actions of, 90–92; infighting in, 107; labor connection of, 86–88, 101–2; legal battles of, 106–7; organizational difficulties of, 101–8; origins of, 85; pooling in, 93; rivalry with other farm groups, 107–8; as success story, 14; and violence, 102–4
National Farmers Union (NFU): and antitrust, 15–16; and bargaining legislation, 97; and corporate farming, 21, 33, 36; *The Corporate Invasion of American Agriculture*, 33; during Nixon era, 15–16
National Land Reform Conference, 31, 35
National Tax Equality Association (NTEA), 116–18, 119, 120, 121, 132, 134, 162
NCFC. *See* National Council of Farmer Cooperatives
Nebraska, 9, 28, 44
Nelson, Gaylord, 22, 23, 25, 27, 33
NFO. *See* National Farmers Organization
NFU. *See* National Farmers Union
Nixon, Richard: and agriculture in 1968 election, 99; and AMPI, 132; and corruption, 23; on meat prices, 47; and the small producer, 15
North Dakota, 7, 9, 32, 79
NTEA. *See* National Tax Equality Association

Obey, David, 22
oligopoly, 1, 4, 5–6, 10, 11
O'Neill, Tip, 157
OPEC, 12
Organization of American Historians, 17
O'Rourke, Edward, 104

Packers and Stockyards Administration, 52
Patton, James, 107, 109, 119, 138, 154
Pearson, Drew, 73
Peek, George, 65, 134
Peritz, Rudoplh, 173

Peterson, Trudy, 69
Public Law 480: and McGovern, 145; operation of, 70; origins of, 69–70
Pocock, J. G. A., ix–x
Pratt, William, 9
Pressler, Larry, 80–81
Publius, 17
Putnam, Robert, xi

Ray, Robert, 78
Ray, Victor, 135
Raymond, C. Elizabeth, 168
Reagan, Ronald, 3, 78
republicanism, ix–xi, 165–67; and corporate farming, 31–32; and farming, 6
Reuther, Walter, 86, 104, 118
Ribicoff, Abraham, 156
right-to-work laws, 54
Robinson, Joan, 49
Rodefeld, Richard, 35–36
Roosevelt, Franklin: farm program, 65
Roosevelt, Theodore, 1
Rowe, Frederick, 12
rural economic development, 8–9
rural outmigration, 7–8; and big cities, 167; and corporate farming, 32
Russian grain sales. *See* Soviet grain sales

Saloutos, Theodore, 137
Sandel, Michael, xi–xii
Sapiro, Aaron, 111
Schlebecker, John T., 47
Schlesinger, Arthur, Jr., xi, 145
Schneidau, R. E., 89
Schultz, George, 74, 76
Schumpeter, Joseph, 2
Sherman Antitrust Act, 3, 16
Simons, Henry, 2
Sommer, Lynnita, 150
Sorenson, Phillip, 35
Sorenson, Theodore, 9
South Dakota: and corporate farming, 21; and rural economic development, 8; rural outmigration in, 7–8
Soviet grain sales, 62, 75–77; and diplomacy, 77–78; and information problems, 78–79; and Presidential politics, 78
soybean industry, 58

Stans, Maurice, 123
Staley, Oren Lee, 34, 81, 86, 88, 91, 154
Stevenson, Adlai, 139
Stigler, George, 13, 105
Strange, Marty, 27
Sweezy, Paul: *Monopoly Capital*, 1–2

Talbot, Ross, 146
TEA (Trade Expansion Act), 72
Temporary National Economic Committee (TNEC), 11, 39
Thatcher, M. W., 20, 23, 32, 114–15, 118, 125, 149, 150
Tintsman, Dale, 56
TNEC (Temporary National Economic Committee), 11, 39
Torgerson, Randall, 96, 131
Trade Expansion Act (TEA), 72
Turner, Daniel, 86–87, 101
Turner, Frederick Jackson, 2
Tweeten, Luther, 176

Udall, Mo, 23
Unger, Douglas: *Leaving the Land*, 19–20
United States Sugar Act, 72
Utley, Garrick, 28

Von's Grocery, 11
Voorhis, Jerry, 134

Wagner, Robert, 157
Wallace, Henry, 117, 124, 137, 157
Walters, Charles, 94, 107
Webster, Daniel, 32
Wellford, Harrison, 29
wheat industry: concentration level of, 59–60; and National Macaroni Manufacturers Association, 60
White House Task Force on Antitrust Policy, 11
Williams, John, 118
Wills, Gary, 167, 170
Wilson, Edmund, 168
Wolf, Fones Elizabeth, 110

Yeager, Mary, 45
Yeutter, Clayton, 100